Bridge to the Rain Forest

Medical missions at the jungle's edge

by

Eleanor Boyes

HCJB
WORLD RADIO

U.S.A. - P.O. Box 39800, Colorado Springs, CO 80949-9800
Canada - 2476 Argentia Road, Suite 201, Mississauga, ON L5N 6M1
Ecuador - Box 17-17-691, Quito, Ecuador, South America

Published by World Radio Missionary Fellowship Inc.

CANADIAN CATALOGUING IN PUBLICATION DATA

Boyes, Eleanor
Bridge to the rain forest

ISBN 0-9699423-0-3

1. Missions, Medical--Ecuador--Shell--History.
I. World Radio Missionary Fellowship. II. Title.
R722.B69 1995 266' - 009866' 4 C95-900268-5

Second printing 2001

Printed in Ecuador

Front Cover: A suspension bridge links the hospital's two complexes. Photo by Beth Huddleston.

Contents

Acknowledgments vi
Preface vii
1 Scarred Faces 1
2 The Stamp of History 11
3 From Swampland to Air Base 16
4 Sharing the Vision 23
5 Medicine Without Doctors 28
6 A Surgical Scratch Team 36
7 A Promise Fulfilled 44
8 The Water Tower 52
9 Launching Epp Memorial Hospital 60
10 Perils on the Road 68
11 Surgery by the Book 78
12 A Shell Mera Romance 86
13 Hospitality 92
14 The Interests of Others 101
15 Disasters 110
16 The Little Ones 120
17 Adversity 126
18 Epidemic in Aucaland 134
19 Triumph in Tragedy 142
20 Mending and Making Do 152
21 "Explosion, Exploration, Eruption!" 159
22 Bridge to the Community 167
23 Battling the Elements 175
24 Perils of the Rain Forest 184
25 Reaching Out 193
26 God's Open Door 201
27 Problems, Provision, Praise 207
28 The New Link 217
Epilogue 227
For Further Reading 230

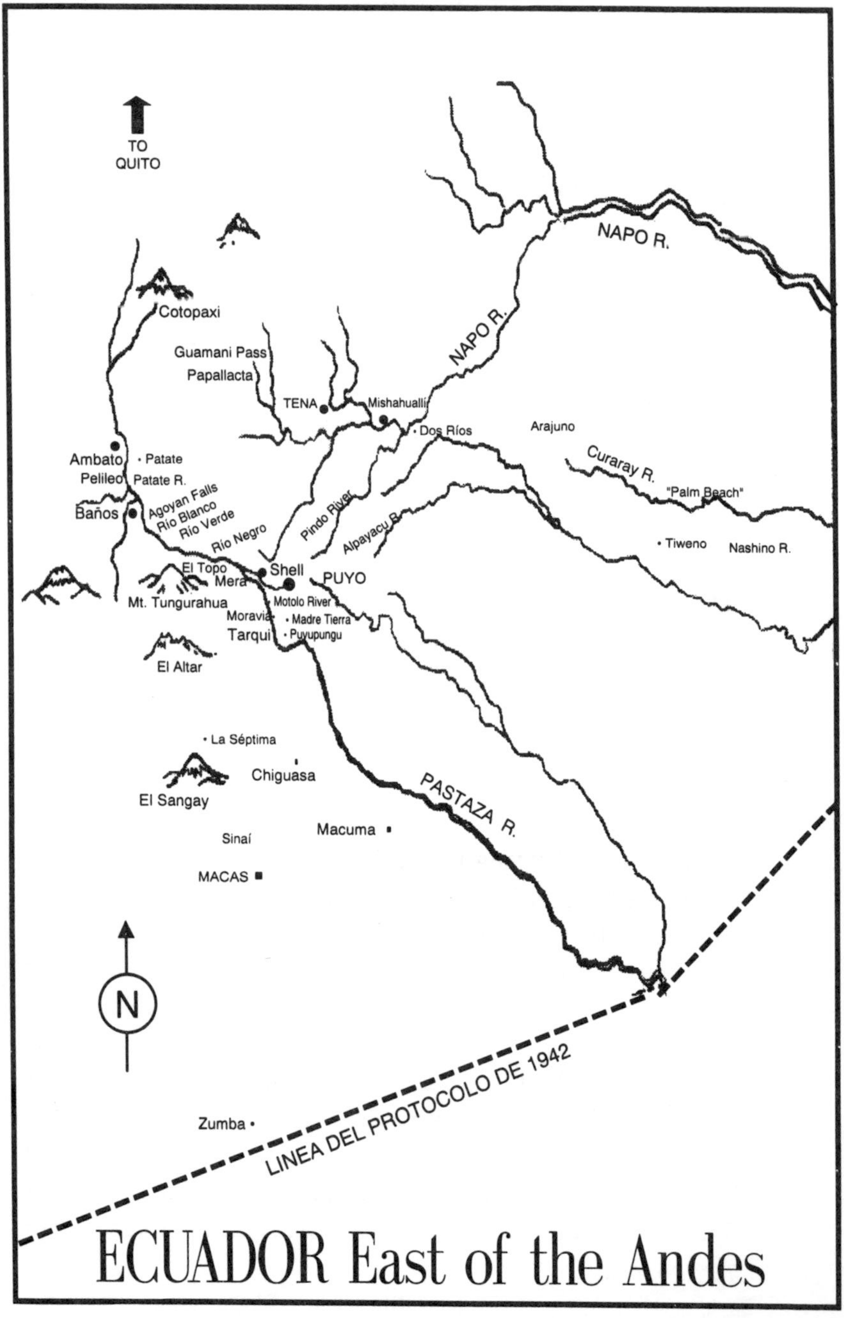
TO
QUITO
Cotopaxi
Guamani Pass
Papallacta
TENA
Mishahuallí
NAPO R.
NAPO R.
Dos Ríos
Arajuno
Curaray R.
"Palm Beach"
Ambato
Patate
Pelileo
Patate R.
Agoyan Falls
Río Blanco
Río Verde
Baños
Río Negro
Pindo River
Alpayacu R.
Tiweno
Nashino R.
El Topo
Shell
Mera
PUYO
Mt. Tungurahua
Motolo River
Moravia
Madre Tierra
Tarqui
Puyupungu
El Altar
La Séptima
Chiguasa
El Sangay
PASTAZA R.
Macuma
Sinaí
MACAS
N
Zumba
LINEA DEL PROTOCOLO DE 1942
ECUADOR East of the Andes

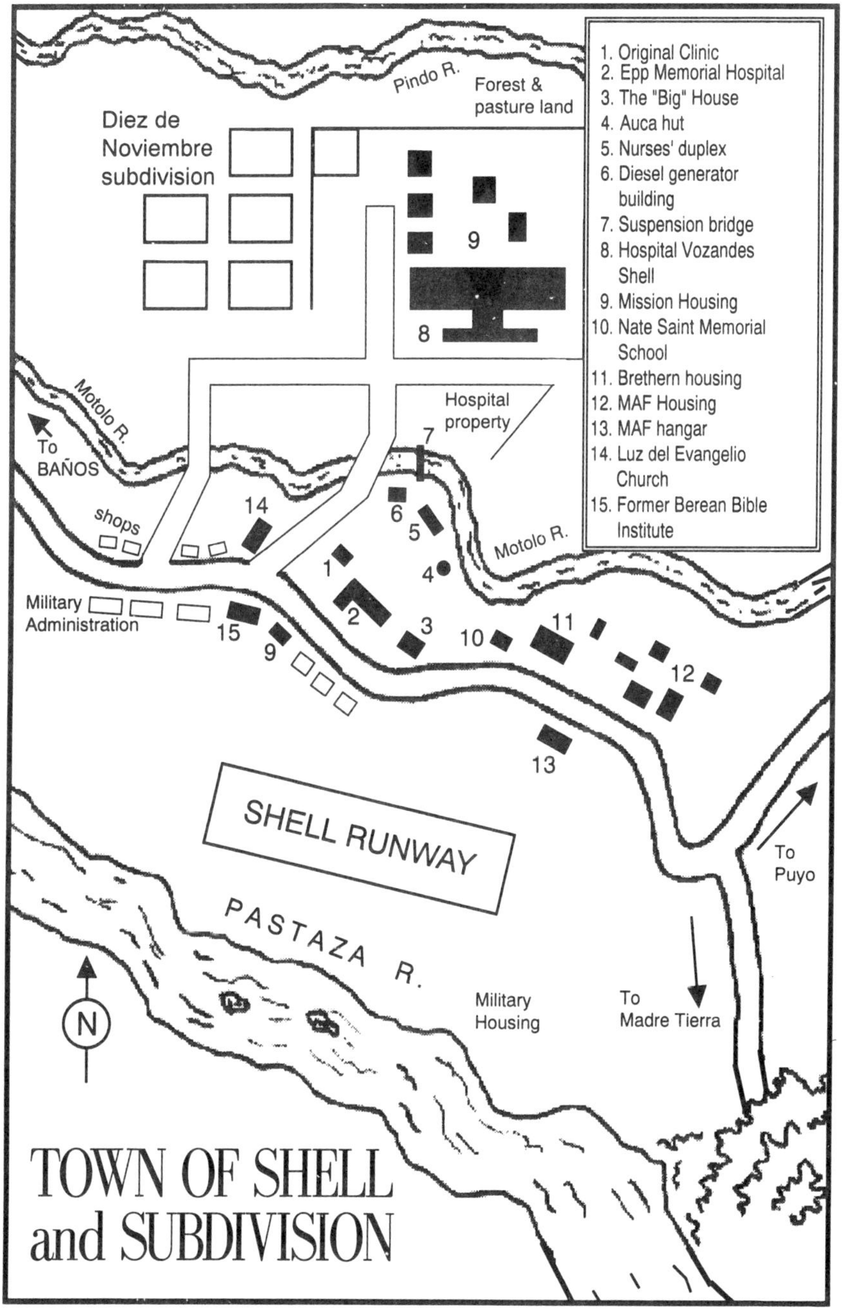

1. Original Clinic
2. Epp Memorial Hospital
3. The "Big" House
4. Auca hut
5. Nurses' duplex
6. Diesel generator building
7. Suspension bridge
8. Hospital Vozandes Shell
9. Mission Housing
10. Nate Saint Memorial School
11. Brethern housing
12. MAF Housing
13. MAF hangar
14. Luz del Evangelio Church
15. Former Berean Bible Institute
Pindo R.
Forest & pasture land
Diez de Noviembre subdivision
Motolo R.
To BAÑOS
shops
Hospital property
Motolo R.
Military Administration
SHELL RUNWAY
PASTAZA R.
Military Housing
To Madre Tierra
To Puyo
N
TOWN OF SHELL and SUBDIVISION

Acknowledgments

I wish to express my appreciation to all the persons who have contributed to this book by providing me with excerpts from their diaries, letters, related stories, and answers to my many questions. You have helped me to put this story together. To those of you who have read the manuscript, in part and in its entirety, I extend my heartfelt thanks. Harold Goerzen, Karen Schmidt, Ruth Ann De Flon and Marilee Dufendach made many helpful suggestions. Robert Hough read several chapters while recuperating from major surgery, and Amory and Norma Shields provided their lovely home by the lake for reading sessions. Special thanks go to Jo Hébert who graciously invested many hours of work to key the manuscript into a computer and, with her helpful editorial skills, gave the manuscript its final polish.

This acknowledgment would not be complete without an expression of my deep appreciation to Beth Huddleston, whose dedication to medical missions and the spread of the gospel has been an inspiration. Without her help and support, this book might not have been written.

Eleanor Boyes

Meaford, Ontario
December 1994

Preface

My earliest recollections of the mission hospital in Shell, at the edge of Ecuador's eastern jungle, are those of a small, unpainted, frame building set against a backdrop of green forest. Devoid of any attempt at contrived "decor" and furnished only with the most basic of hospital equipment, the building's interior seemed stark. White bed linen in the patients' rooms contrasted sharply with the dark wooden walls. At night a family member slept on the floor beside each patient's bed, ready to attend to the needs of their sick relative. Indeed, the mission hospital blended well with its surroundings, reflecting the primitive conditions of the area.

In those early days, as I remember them, missionaries assigned to Shell seldom made the long, tiring trip to Ecuador's capital city of Quito. This meant that they had virtually no social life, for at Shell electricity, a luxury, was cut off abruptly at nine o'clock each night when the hospital's diesel generator shut down. As the cut-off hour approached, everyone made a dash for bed, for the blackness that descended so suddenly was overpowering. The drone of countless insects vibrated through the tropical night air, replacing the sound of daytime traffic, since after dark the nearby road was deserted and silent.

Years later, when I joined the staff in Shell and listened to the stories of bygone days told by my colleagues, both foreign and Ecuadorian, I decided that many of their recollections should be recorded. If history were to be preserved, it must be written; otherwise it would soon be forgotten.

I have written this book not only for those who are currently serving in the medical ministry in Shell, but also for the great host of people who have become interested in their work through visiting the area or through contact with the missionaries there. I have also written this book for those who will one day serve the Lord in medical missions. It is my hope that the reader will see the hand of God working in the lives of those who have served Him there at the edge of Ecuador's rain forest.

Although the persons and places mentioned in these pages are largely real, I have changed some of the Ecuadorian names. Over the years, some names, though not the circumstances surrounding them, have been forgotten, and to heighten the reader's sense of reality, I have taken the liberty of giving names to those people. Names of foreign persons, however, unless it is so stated, have not been changed. Nor has it been possible to record all the events or the names of all the individuals who, down the years, have been involved in the work at Shell. To those people who have served there, and whose names do not appear in this record, I apologize. But God has recorded your achievements in His "book of remembrance," and to Him we give the praise and the glory.

Eleanor Boyes,
Meaford, Ontario
December 1994

I
Scarred Faces

"Man looks at the outward appearance, but the Lord looks at the heart."

–1 Samuel 16:7b

In the wooded ravine behind *Hospital Vozandes del Oriente*, the full moon, huge and orange, slipped silently over the treetops, peeping through the branches and sending deep shadows across the thick green carpet of turf. With the moonrise, the dark patches on the hospital lawn gradually receded. As the moon rose above the world's largest rain forest, its brilliant glow softened from deep orange to yellow and then to ivory.

The peaceful scene outside the mission hospital was reflected inside its walls. With the patients settled for the night and no emergencies or maternity cases to delay the staff, it appeared that the evening shift would finish work at 11 o'clock. Gloria, the nurse's aide in charge, was a short, chunky, dark-haired woman with more than ten years of nursing experience. Only four feet, eight inches tall, she stood to do her desk work. Her assistant, Lidia, was the nurse's aide responsible for direct patient care.

It was late when Dr. Roberto, the Ecuadorian physician on duty, arrived at the nurses' station with the previous day's newspaper tucked under his arm. He sat down to read and soon became engrossed in the editorial section of *El Comercio*, Ecuador's leading newspaper. The paper usually reached the hospital a day late, but between it and his transistor radio, the doctor kept himself well informed of daily happenings within the country. He would not retire for the night until he had read *El Comercio* from front to back.

Lidia entered the little kitchen opposite the nurses' station to prepare the staff's evening snack. Placing a pan of milk on the gas stove, she lit the burner. The milk was for the *café con leche* (coffee with milk), preferred by most Ecuadorians. On busy evenings it almost always boiled over, extinguishing the burner's flame. It would do so again tonight, for at that moment the emergency bell rang. Its shrill sound vibrated throughout the wooden building, sending the medical staff into action.

Lidia was the first to scurry along the corridor toward the main entrance. Roberto dropped his paper and followed.

"I have five people in my cab who've been badly burned in a house fire!" shouted an excited taxi driver as he waited impatiently for Lidia to unlock the door. He continued to explain as he hurried the nurse and doctor down the cement ramp to his parked car. After listening to the agonized cries of his passengers for more than an hour, the driver was finding it difficult to control his emotions. Trembling, he wiped the perspiration from his brow with his shirtsleeve and begged, "Doctor, please take care of them!"

Glancing inside the taxi, Roberto and Lidia saw Dolores Espín with her three daughters and one son, shivering and crying with intense pain. "*¡Por favor, doctor, por Dios, ayúdanos!*" ("Please, doctor, in the name of the Lord, help us!") Dolores pleaded at her first glimpse of the doctor's face. The nauseating odor of singed flesh filled the cab. Lidia turned and ran up the ramp into the hospital for stretchers.

The Espín family was from San Francisco, a hamlet midway between the towns of Shell and Baños. Most travelers to Shell connect San Francisco with a nearby waterfall, which cascades down a rocky wall, spills across a stony road, and drops over a cliff to join the Pastaza River in the gorge below. On the evening of the fire, the Espín family had planned to celebrate the children's success in passing their school exams. While Dolores prepared a special meal, her husband, Oswaldo, was celebrating with his buddies at the local saloon. Dolores did not expect him home for several hours.

The children had spent the afternoon playing outside their house. When the sun set behind the steep Andean peaks, Dolores called them inside and padlocked the door. Then one of the children lit the only lamp the family owned. It flickered and went out for lack of fuel. In the dim light, a child, no one remembers which, picked up the can of gasoline and filled the kerosene lamp. When she struck the match, the gas exploded, and flames leaped through the lightly built frame home. The family panicked; no one could find the key to the padlock. Desperate to escape the flames, they managed to climb through the only window.

Alerted by screams, neighbors rushed to the scene to offer help, but flames quickly consumed the wooden house. Seeing the family's urgent need of medical care, the would-be helpers quickly turned their attention to the burn victims. They insisted that the only taxi driver in the hamlet take the Espíns to the mission hospital, an hour's drive across a road consisting of sharp mountain curves.

The mission hospital was an unpretentious L-shaped structure on the main thoroughfare at Shell's southeast end. Located on the fringes of the eastern jungle, Shell is a five-hour drive from Ecuador's capital city of Quito. Easily seen from the road and illuminated by a fluorescent light at night stood a sign reading *HOSPITAL VOZANDES DEL ORIENTE*, (*Vozandes* is a contraction for *La Voz de Los Andes*, or The Voice of the Andes). *Oriente* signifies the eastern provinces of Ecuador's vast forest area of the upper Amazon basin. The hospital is owned and operated by World Radio Missionary Fellowship, Inc., often referred to as HCJB World Radio, the world's pioneer missionary broadcaster.

By North American standards, this sparsely equipped mission hospital lacked up-to-date technology. Yet the local people had found that its staff administered prompt and effective medical attention. Sometimes these people went elsewhere for minor treatment, but they brought to the mission hospital all their relatives who were seriously ill.

At the emergency entrance, Roberto quickly took command of the situation. By this time, in August 1983, the mission hospital was no longer the only medical facility in the area. With a capacity of 25 patients, it had neither staff nor equipment to treat several emergency cases that would demand many days of intensive nursing care. Without hesitation the doctor turned to the taxi driver and said, "We'll keep the two most critical patients. You take the others on to the military hospital here in town."

The two patients with the most extensive burns were nine-year-old Margot and seven-year-old Silvia, who had badly burned faces. With the help of the taxi driver, Roberto and Lidia placed the two sisters on stretchers, covering their half-naked bodies with blankets. The little girls shrieked with pain and fear as the staff wheeled them up the ramp into the emergency room.

Since Silvia was the more severely burned, the doctor turned first to her. Taking a plastic intravenous (IV) bottle from a nearby shelf, he removed its pointed tip, connected the tubing, and hung it upside down on a hooked wire dangling from the ceiling. (The improvised wire hooks replaced portable IV poles that occupied too much floor space.) Skillfully he inserted a delicate intravenous needle into the only unaffected vein in Silvia's forearm. The liquid would replenish the fluid the child had lost through burns. Meanwhile, trying to avoid touching the burned areas of their bodies, Lidia began to undress the little girls.

At the nurses' station Gloria heard the children's screams.

Curious, she hurried down the hall, opened the emergency room door a crack, and peered in. "Bring 50 milligrams of Demerol for Margot, Gloria, and 35 milligrams for Silvia," Roberto ordered.

As quickly as she could, Gloria brought and administered the pain-relieving drug to the two sisters. Behind dark-rimmed glasses, Roberto's brown eyes registered concern. "Please notify the physician on second call," he said.

Gloria was just closing the door when Roberto spoke again. Pointing with his chin as he glanced in Lidia's direction, he added, "Call another nurse. Lidia needs some help, too."

Since Australian nurse Wendy Freeman was on call, Gloria dialed her number. Wendy responded at once: she would be right over. The second doctor on call, *Doctora* Alma, was attending a women's meeting at the evangelical church, one block up the road. Since the church had no telephone, Gloria slipped out through the hospital's front door and ran up the road. She paused, breathless, and then opened the building's squeaky door. All eyes, including Alma's, turned toward her. At Gloria's frantic signal, Alma hurried from the room, and together they rushed back to the hospital. Alma's short, quick steps and the click of her high-heeled shoes sounded sharply as she rushed up the ramp into the emergency room.

Inside, Wendy was helping Lidia remove Silvia's dress. The bits that remained of the child's singed garment stuck to her body like glue. With scissors, Wendy cut along the seams and then, as gently as possible, she and Lidia peeled the blackened fabric from Silvia's shivering body. They quickly covered her with a sheet and blanket before moving over to Margot's stretcher to do the same. The sisters screamed each time someone touched them. A hospital was a new experience for both children, and separation from their family increased their fear of the unknown.

Dressed in sterile gowns and wearing surgical gloves, the two doctors meticulously cleansed the burned areas of the children's bodies with antiseptic solution. They gently swabbed the burned flesh, washing and peeling away the dead tissue as one strips a room of old wallpaper.

An hour later, Kathy Jo Estes, a nurse and mother of four, walked home from the women's meeting. Why, she wondered, were the lights still on in the emergency room? She decided to check.

On entering thc hospital, Kathy Jo covered her street clothes with a white hospital gown before entering the emergency area. The room was in disarray: dirty instruments and used surgical gloves lay in the sink; wrappers from sterile equipment were strewn everywhere, and the hamper provided for soiled linen overflowed. To make space, the staff had moved Silvia to the table and pushed her cart into the corridor. Margot's stretcher remained squeezed close to Silvia.

"Oh, for the new hospital!" Kathy Jo breathed to herself. The new hospital, already being constructed, would provide ample space to deal with such emergencies. Aloud, Kathy Jo said to Lidia, "Can I help you?"

Lidia nodded. Wendy and the two doctors were keeping her busy with their repeated requests for more saline solution and gauze. Lidia needed more supplies and was anxious to replenish them. Thankful for Kathy Jo's help, she scooped up an armful of empty containers from the counter and headed for the central supply room.

It was nearly 11 p.m. when the doctors finished their work. Together with the nurses they wheeled the two sisters along the corridor to the inpatient section. To prevent cross-contamination, they placed Margot and Silvia between sterile sheets in improvised burn tents, isolating them in a room separate from other patients. This would be their environment for the next several days.

Two hours had passed since Kathy Jo had entered the hospital. Leaving through the main entrance, she headed for her home, across the road. Overhead, in a cloudless sky, the full moon bathed the scenery with silver splendor. Outlined against the western horizon was the eastern range of the Andes Mountains. Moonbeams sparkled on two snowcapped peaks. One was Sangay, an active volcano that rises 17,000 feet above sea level, 30 miles south of Shell. Normally seen belching clouds of smoke from its cone tip, Sangay was inactive that night. Farther up the range, the jagged snow-covered summit of *El Altar* reached heavenward, illustrating by its shape the name it bore. This majestic mountain resembled an altar in some great cathedral. Someone had appropriately named its peaks: Bishop, Friar, Tabernacle, Canon and Nun.

Kathy Jo paused for a moment to take in the splendor of the panorama before her. Visitors seldom witnessed this breathtaking beauty reserved for those who live at the foothills of the Andes. "How wonderful of the Creator of the universe to arrange a medical work in such a beautiful setting!" mused Kathy Jo. As she slipped silently into the house, she breathed a prayer of thanksgiving for the privilege of living and working at the jungle's threshold.

The quiet moonlit night was in sharp contrast to the activities inside the mission hospital. The emergency treatment of the two sisters had occupied the evening staff so long that the next shift was now arriving. Judy Cole, a missionary nurse from England, and her helper, Maria, were working the midnight shift. After punching their time cards, they busily counted narcotics, syringes, baby bottles, stethoscopes, and the change in the money drawer. Then they sat down at the desk to listen to the evening report.

Gloria was halfway through her report when the buzzer at the front door rang. Judy and Maria exchanged glances; they knew that they were in for a busy shift. This sudden change from a peaceful evening to a busy one was typical of medical work at the jungle's edge. Either the hospital was so quiet that the nurses were bored or so busy that staffing was inadequate.

The next moment, Roberto and Lidia hurried past the desk, pushing a gurney on which a pregnant woman lay screaming, "*¡Ay, ay, ay! Ya no aguanto más!*" ("Oooooooh! I can't bear any more!") Without pausing, they wheeled the wailing woman into the delivery room.

Gloria was resuming her unfinished report when a patient in Room 11 rang his bell, a prearranged signal that his IV bottle was nearly empty. Another interruption! Judy looked at María and sighed. Though she did not utter a word, María understood; she, too, wondered when the previous shift's report would be finished.

Gloria picked up the new IV bottle and headed for Room 11 just as a barefoot Indian mother carrying a whimpering child from the pediatric ward appeared at the desk. Unable to speak Spanish, the woman held up an empty baby's bottle. María took the empty one and went to the kitchen to replace it. Judy gathered the tickets for the medicines to be given at midnight. She was checking them when Gloria returned. Finally, Gloria completed her report.

It was after midnight when Gloria and Lidia stopped work. Weary, they punched their time cards. In an incubator at the nurses' station, the hospital's newest arrival kicked furiously while he moved his lips back and forth in sucking motions. Dr. Roberto, on call for the remainder of the night, sat writing orders on the new admission. "I wonder," Gloria whispered to Lidia, "if he'll find time to finish reading his newspaper."

Down the hall, Judy and Maria entered Margot and Silvia's room. They found both girls awake and restless. Margot, with the thirst typical of burn patients, was trying to drink from a bottle of antiseptic solution inadvertently left on her bedside table. Her intravenous needle had slipped out of the vein and was leaking fluid into the surrounding tissue. Judy removed the needle and applied pressure to the area. Now the doctor must look for a new vein.

Usually a relative stayed with any critically ill or anxious patient because the presence of a loved one helped to restore calm. Margot and Silvia had no family member to stay with them, for Oswaldo Espín did not learn of the tragedy until he wandered home from the saloon at midnight. Finding his home in ashes and frantic for news of his family, Oswaldo set out at once to search for them. Because the Espíns were civilians, staff at the military hospital had sent them on to healthcare facilities in Puyo, capital of Pastaza Province, five miles east of Shell. When Oswaldo reached Puyo, he found his wife and son in the same hospital; they would make a good recovery, but a daughter admitted to another hospital, had died in surgery. Distraught, he returned to Shell. At the mission hospital he asked to see Margot and Silvia. Though he was relieved to find them alive, his eyes filled with tears when he saw their burns.

The next few weeks were stressful for the little sisters. Each day a doctor debrided their burns, removing the dead skin and crusts and cleansing the affected areas with a germicidal solution to help prevent infection. The treatment was painful, for raw flesh smarted under the pressure of the doctor's antiseptic swab. Just as distressing and frightening were the daily "burn" baths.

The care of the two children was a physical and emotional drain on the nursing staff, who found it hard to look at their disfigured faces. "Will Silvia and Margot be left with permanent scars?" they asked one another. "Will surgeons ever be able to restore their facial features? Will their playmates accept them? Will they feel rejected because they are disfigured?"

Finally the day came when Margot and Silvia were released to go home. But several months later they returned to Shell for further treatment. This time the surgeon removed scar tissue to relieve the thickened and shortened muscles behind their knees and at their elbows. Then he grafted skin on these areas so that the children could again extend their arms and legs.

Despite these operations, Margot and Silvia would never look the same as they did before the fire.

During their hospital stay, the chaplain visited the little girls often, explaining God's love for them. Gradually, the seed he planted began to take root. One day the Lord used the Ecuadorian nursing director to bring Silvia to the Great Physician, the One whom she learned would never reject her scarred face, but loves her as she is. Later, when Margot and her mother understood the Savior's love for them, they, too, accepted Him. They all know now that God does not look on Margot's and Silvia's outward appearance, but on their heart, where lives the joy that He has given them.

2
The Stamp of History

"He who oppresses the poor shows contempt for their Maker."

-Proverbs 14:31a

On the night when Margot and Silvia arrived for burn treatment at the emergency entrance of *Hospital Vozandes del Oriente*, the mission hospital had completed 25 years of service. Day and night its doors had opened to all sorts of persons--rich and poor, from far and near--with all manner of diseases and injuries. The hospital's reputation for compassionate care by qualified medical personnel had spread throughout the *oriente*, an amazing development given the history of missionary work in the region. Not only were the Indians hostile to the "outsiders," but also travel in the *oriente* was slow and arduous.

Reuben and Grace Larson were the first Protestant missionaries to take the gospel to the isolated people of Ecuador's eastern region. The Larsons, pioneers with the Christian and Missionary Alliance (C&MA), arrived in Ecuador in 1924, spending their first two years at the coast, where they studied Spanish and distributed Bibles. When they moved to the *oriente*, they settled at the confluence of the Tena and Mishahualli Rivers, a place known as Dos Ríos (Two Rivers).

Reuben Larson took with him into the jungle a number of items unobtainable there and, with this stock, set up a trading post to attract the Quichua Indians. He had hoped to entice some of these people to work for him in exchange for matches, fishhooks, salt, knives, and handkerchiefs, and thus create opportunities for bringing them the gospel message. Although his plan seemed a good one, it failed to work. With a few coins, the Indians would buy an item in Larsons' store and then leave, giving the newcomers no opportunity to talk with them. The people the Larsons had come to evangelize avoided them.

The Quichuas, whose ancestors had suffered deeply at the hands of the *conquistadores* during the Spanish invasion of 1532, trace their ancestry back to the Inca empire. Under the leadership of an adventurer named Francisco Pizarro, Spanish conquerors hacked their way through unexplored forests in search of gold. Considering themselves superior to the Quichuas and determined to enforce their authority, the *conquistadores* plundered and tortured as they went. Some of Pizzaro's men died of starvation in the rain forests, and others of wounds inflicted by the Indians' poison darts. Still, in terror of the Spaniards with their "modern" weapons, the mountain people fled higher into the cold bleak slopes of the Andes, while the Amazonian tribes slipped silently into the dense jungle, protected from the foreigners by inaccessible rain forests. Stories of cruelty, handed down from one generation to another, instilled in the Indians a fear of Europeans.

The Spanish conquest changed not only the political organization of South America, but also its ethnic balance. As Spain colonized its lands, Spanish men took tribal women for their pleasure, fathering the first generation of *mestizos* (persons of mixed race), and thus reducing the number of pure-blooded Indians and Spaniards. Next to invade the *oriente* were explorers from the world's industrialized countries. In search of rubber trees, oil, gold and other forms of wealth, they, too, ruthlessly exploited the people. Generations later the native people still feared foreigners and wanted nothing to do with them. It was into this hostile environment that Reuben and Grace Larson came.

The Larsons spent several months in the jungle without making any perceptible progress in presenting the gospel. By this time they knew that if they were to evangelize the Indians, they could not do so by human power. Had they, they wondered, relied too much on their own strength? One Sunday, broken in spirit, they knelt in their home and asked God for His help. As they prayed, God gave them the answer: "Not by might nor by power, but by my Spirit,' says the Lord Almighty" (Zechariah 4:6).

The following day 40 Indians in two dugout canoes paddled upriver to visit the Larsons. Their chief, Domingo Cerda, who was a witch doctor, stood before Reuben and Grace, dug his chonta-wood pole into the ground, spat--the preliminary gesture to any speech--and said, "The other foreigners told us that you were servants of the devil. They said that the stone at the door of your house was *Señora* Rosario, the woman who disappeared some months ago. They told us that you turned her into stone, and that you stomp on her every time you go into your house. But we've been watching you. If you are servants of the devil, we want to serve him, too."

Selfish white settlers who wanted to keep the territory for themselves had spread these ugly tales about the Larsons. But in spite of their efforts to prevent the gospel from reaching the Indians, the door had now opened for the Larsons to evangelize.

If contacting the local people was difficult, traveling to and from the *oriente* was just as hard. To reach Dos Ríos from Ecuador's capital, a 12-day journey sloshing through mud, the Larsons had a choice of two routes. They could either cross the cold, misty, 13,000-foot Guamani Pass or take the longer but easier route across the mountains, through what has now become the town of Shell.

The Larsons used various modes of travel on the trip through Mera, Shell and Puyo. First they traveled south to Ambato by train; then east on horseback along a shelf-like path that zigzagged through the mountains. This trail followed the Pastaza River through a narrow, winding gorge. Slogging through the rain and mud, their horses had been known to sink to their bellies, and in the frustration of trying to extricate themselves, would lunge from side to side, sometimes throwing off both rider and baggage.

At the village of Mera, the Larsons left the horses and hired Indian carriers, whose knowledge of the thick forest is indispensable to the foreigner. The Indians led the missionaries on foot through the area where Shell would be located, halfway between Mera and Puyo. As the Larsons made this long journey from Quito to their jungle home at Dos Ríos, Reuben viewed the *oriente* of the 1920s as a dense, impenetrable rain forest; the uninhabited region of Shell seemed nothing more than worthless swampland with porous soil. Little did he know that in years to come, he would see a mission hospital in this area.

From Shell, the Larsons traveled on to Puyo, a hamlet consisting of no more than eight or ten thatched huts. Beyond Puyo the trail ended, and from there the Larsons traveled by canoe. At sundown the carriers built temporary lean-tos, using poles, branches, leaves and twigs. The abundant jungle growth provided plentiful foliage for both roof and floor; a rubber poncho laid over the leaves protected the Larsons from dampness.

The Ecuadorian jungle--hot and humid in the daytime--is cool at night. Before crawling into sleeping bags, the Larsons removed their wet clothes and put on dry ones. The next morning they struggled into soggy socks, wet boots and damp garments, tucking their dry togs into a weatherproof bag to wear the following night. Because of the constant dampness, they often had to contend with mold and mildew. On one trip, it rained for six straight days--the sun never came out. When Reuben and Grace finally reached their destination, Reuben removed his hat to find it coated with mildew.

On these journeys, the Indians took their food with them.

Their main nourishment is *chicha*, masticated *yucca* (manioc), first boiled and then chewed by the women, who spit the mixture into a gourd. (Saliva, a yeast agent, starts fermentation.) The women wrapped the mash, called *huito*, into leaf packages which the men took with them and buried at different places.

On their return trip they dug up the packages, washed away the dirt that clung to them, put the stringy, lumpy mash into a bowl of water and drank the mixture.

At home, in Dos Ríos, Grace Larson discovered that the fermented *chicha* served other purposes as well. She employed a Quichua girl, Antonia, to help her with the housework. Using an adobe oven, Grace taught Antonia the art of breadmaking, but soon Antonia was producing better bread than Grace. One day Grace asked, "What's the secret of your good bread, Antonia?"

"Oh, *Señora*, it's the *chicha*!" Antonia answered proudly.

The Larsons spent several years working in the *oriente* with their beloved Indians. During those years, Reuben often considered the possibility of using radio to spread the gospel. On furlough in the United States, he met Clarence Jones, a young man keenly interested in beginning a radio work in South America. Before returning to Ecuador, Reuben and Clarence drew up a contract that Reuben and a missionary colleague, Stuart Clark, presented to the Ecuadorian government. Ecuador's congress, wishing to see the country develop, granted the missionaries a 25-year contract. On Christmas Day, 1931 Radio Station HCJB, whose call letters stand for Heralding Christ Jesus' Blessings, began broadcasts. As the station added programs in more languages, broadcasts in the Indian dialect became a valuable tool to reach both lowland and mountain Quichuas with the gospel.

3
From Swampland to Air Base

"...what has happened to me has really served to advance the gospel."

-Philippians 1:12

More than a decade after the Larsons made their first trip into the *oriente*, geologists working for the Shell Oil Company came to Ecuador in search of petroleum. Owned by the Dutch and operated by the British, Shell Oil sought permission from the Ecuadorian government to begin exploring in the jungle lowlands east of the Andes Mountains. Since the Ecuadorian government lacked the huge sums of money required to build roads in that region, it happily granted the company permission to proceed with its explorations throughout the whole virgin jungle area.

Good geologists, it is said, are prospectors at heart, for they love exploring. Yet Shell Oil surveyors soon found that the geographical knowledge at their disposal was limited. Moreover, the Andes Mountains, with their steep slopes and deep valleys, complicated transportation and kept conditions primitive in the isolated *oriente*. The region had footpaths, but no roads, no airstrips and few friendly Indians.

Shell Oil faced a mammoth task in attempting to open up the *oriente*. Before 1938, government representatives and missionaries traveling along the primitive trails to the *oriente* contended with frequent rock and mud slides caused by the heavy rainfalls. Only by crossing raging rivers on handmade cable cars were exhausted travelers finally able to reach their destinations.

Undaunted, Shell Oil's surveyors, searching for a suitable site on which to construct a base camp and airstrip, made the difficult journey around mountainous slopes and through the Pastaza River gorge to the village of Mera. Finding the area about Mera, with its hilly terrain, unsuitable for their purposes, the surveyors pushed on. When they arrived at the confluence of the Alpayacu and Pastaza Rivers, they discovered a shoreline that offered an abundance of stone and rock, materials necessary for the construction of a road and airstrip. Leaving the junction of the two rivers and climbing a steep hill, they found a level area of desolate bog land. It was on this stretch of flat land lying between the Pastaza and Motolo Rivers, five miles east of Mera, that Shell Oil decided to build a large air base.

To implement the company's decision, the Shell team's first task was to build a road into the area. The labor required was enormous. Still, using the most modern machinery of the 1930s, Shell's workers blasted through granite and dug through mud and loose gravel to build a road from Baños to Shell Mera. That road, carved out of rugged mountainous terrain, followed much of the old Indian trail that wound through the Pastaza River gorge. For many years, it would provide the only overland access to the *oriente*.

The company named its new settlement Shell Mera, after Shell Oil and Juan León Mera, author of Ecuador's national anthem. There the company built administrative offices and housing for approximately twenty British officials and their families. To construct the solid airstrip Shell Oil needed on top of the area's boggy soil, workers laid down tons of crushed rock. The completed airstrip, one mile in length, ran parallel to the main road. The company's pilots, using four or five airplanes, flew supplies and materials in and out of Shell Mera. Next, to facilitate its exploration for oil, the team built more airfields that are still considered some of the best in the *oriente*. It is doubtful, however, if Shell Oil's officials realized all the hazards their staff would face in their search for petroleum. The risks included poisonous snakes, tropical diseases, and potentially dangerous neighbors, for Shell Oil's first camp, built at Arajuno, was not far from the hostile Auca (Waorani) Indians, who twice speared company workers to death.

Though Shell Oil drilled exploratory wells in several locations, their engineers did not discover sufficient oil to warrant commercial drilling. In 1949, therefore, disappointed by 11 years of fruitless exploration and construction, taxed by the difficulties of jungle living, and poorer by millions of dollars, Shell Oil decided to leave Ecuador. Before the company shut down its operation there, however, two tragic events occurred.

Early in the afternoon of Friday, August 5, 1949 a violent earthquake rocked Ecuador. Severe tremors destroyed the town of Pelileo, 65 miles north of Shell Mera, near Ambato. The earthquake struck within the shadow of Tungurahua, a volcanic mountain that 150 years earlier had buried the town of Baños with lava. Surviving witnesses of the 1949 disaster told how the earth split open and sucked dozens of people into its cracks, then closed, crushing its victims to death. Panic-stricken inhabitants rushed into the Roman Catholic church to pray for forgiveness, only to be swallowed alive, minutes later by another quake. Six thousand people died, and many were injured. Countless survivors were homeless.

Across Ecuador people were desperate for news of relatives living in the disaster area. Shell Mera's principal means of communication with the outside world was the road, but the earthquake had made much of that road impassable. Nevertheless, word of the disaster trickled into Shell Mera, brought by a few survivors who walked across the barriers thrown up by the landslides. Their accounts of the destruction only heightened people's anxiety for relatives and friends.

Anxious to help, Shell Oil offered to fly its employees to the disaster area to search for missing relatives. The day following the quake, therefore, the pilot of the company's big Bristol aircraft took off with a load of passengers and headed toward the mountains, planning to land at Ambato's airport. On the ground at Pelileo, rescue workers saw the plane circling for an aerial view of the disaster area. Then, as the pilot tried to gain altitude, the plane veered westward and was soon lost to sight in thick clouds. Flying through the cloud cover, the aircraft struck the peak of Mount Bolivar, 25 miles west of Tungurahua. All 37 passengers, as well as the British pilot, died in the crash. Even though a search party left immediately, it took several days for its members to reach the site of the crash. Meanwhile, heavy rains had covered the bodies with mud, making identification difficult.

Segundo Aliaga, father of Lola, who years later would become a nurse's aide at the mission hospital, had walked the five miles from Mera to arrive at Shell Mera's runway minutes before take off. Finding a long line of people waiting for the plane, he feared that he would be unable to reach Ambato to search for his wife, who was hospitalized there. When the passenger ahead of him dropped out, however, Segundo obtained a seat on the plane. At home in Mera, five-year-old Lola and her eighteen-year-old sister María Elena were devastated when they received news of the plane crash. María Elena became hysterical, certain that she and her sister had lost both parents. Several days later, however, their mother returned to Shell Mera. She had survived the dreadful earthquake.

Shell Oil's workers must have felt defeated. They had not found enough oil to warrant the building of a pipeline across the Andes to the coast. Now the plane crash, instead of helping the people of Ecuador, had only added further grief to a national disaster. Yet Shell Oil's contribution in Ecuador was highly significant. First, it opened the previously inaccessible *oriente* to Ecuadorians from the western regions. Second, it paved the way for the advance of mission work in the *oriente*. The airstrip at Shell Mera became an excellent landing field, shortening missionary travel in and out of the jungle from a tramp of many days along rugged trails to a flight that took only minutes. Mission organizations already serving in the area could now expand their work, and more groups could join in the task of making Christ known to the people of the rain forest.

Gospel Missionary Union (GMU) was the first mission to enter Ecuador and to establish a work in the southeastern jungle among the Jívaros, now called Shuar, a tribe known for its ancient practice of head shrinking. During the 1940s, GMU owned a plane that served its missionaries in remote areas. But in 1947, an engine fault brought the plane down in thick jungle. In the crash, pilot Bob Hart fractured his knee and ankle. His passenger, HCJB missionary George Poole, also a pilot, suffered minor injuries. Lacking training in radio communication and aeromechanics, the two men were helpless. George, who had not planned to leave Bob alone for more than a few hours while he went looking for the Pastaza River, lost his way in thick underbrush that hid his tracks. Unable to find his way back to the damaged plane, he wandered for nine days without food. By the time he spotted an Indian who led him to Shell Mera, George had lost 40 pounds. Two days later, only a day's trek from Shell Mera, a search team of Indians and missionaries found Bob still alive. He had survived on jungle berries and two raw squash left in the wrecked plane.

The *oriente*, which comprises a third of Ecuador's land mass, consists of an estimated 35,000 square miles of thick rain forest. When a plane is forced to land in this vast area, the jungle treetops separate on impact, and then close again, concealing the damaged plane from pilots making an air search. After GMU lost its airplane, Mission Aviation Fellowship (MAF) began its work in Ecuador. Details of GMU's plane crash served as a caution to MAF leaders to choose their pilots carefully. A survey of the eastern jungles convinced them that they must find a pilot mechanic to launch their work in Ecuador. MAF chose Nate Saint, who became one of the most highly skilled and innovative pilot mechanics in the history of missionary aviation.

Nate and his wife, Marj, arrived in Ecuador in September 1948, one year before Shell Oil's departure from the region. Eager to help the isolated missionaries, Nate began at once to fly supplies and equipment into the jungle stations. On rainy days, when he could not fly safely, Nate worked at building a house and hangar on land opposite the east end of Shell Oil's Shell Mera headquarters and airfield. Meanwhile, he and Marj lived in a tent.

The freshly cut timber that Nate had to use for his building was wet. Nate knew that Reuben Larson, who had built his house with wet lumber, had to contend with cracks in the floors, and so Nate only tapped nails into the floorboards and waited for the boards to shrink. When they had done so, he removed the nails, moved the boards together to eliminate the cracks that shrinkage had created between them, and finally pounded them into place.

In December, Hobey Lowrance of MAF joined the Saints. Hobey, a former American Airlines captain, had obtained permission from his director to spend time in Ecuador to observe first-hand the use of aircraft flying across vast reaches of the Amazon basin. Hobey had been in Ecuador only two days when Nate's yellow Stinson was caught in tricky air currents flowing from the mountains surrounding Quito. In the crash that followed, Nate fractured his back and an ankle. While Nate, in a body cast, was grounded for several months, Hobey did the flying.

During his months in Ecuador, Hobey met and married Olivia Scott, a missionary serving with the Christian and Missionary Alliance. The young couple set their wedding day for August 9,1949. On August 5, however, the earthquake struck, throwing the whole country into upheaval, and upsetting the couple's wedding plans. Hobey and Olivia had asked Stuart Clark to marry them, and Nate was to be the groom's best man. In the wake of the earthquake, Ecuadorian President Galo Plaza asked Radio Station HCJB for help, personally requesting that Stuart Clark coordinate the relief effort. Stuart left immediately for the disaster area. Moreover, with the Shell Mera road impassable, Nate was unable to travel to Quito. Although a quick change of plans was necessary, Hobey and Olivia found substitutes before August 9, and the wedding took place on schedule.

With the help of the engineering department of Radio Station HCJB, Nate and Hobey set up a shortwave radio network, connecting all the jungle mission stations to the Shell Mera base. Each morning the missionaries communicated with MAF. They ordered medicines for the sick, consulted with a doctor in Quito by ham radio, and in a crisis requested an emergency flight.

Overnight, MAF became a bridge by which the isolated jungle people could reach the medical assistance they so desperately needed. Each time the MAF pilot arrived at a jungle station, Indians suffering from various illnesses begged to be flown out for medical help. To provide them with adequate health care now became Nate Saint's dream.

4
Sharing the Vision

"Have I not wept for those in trouble? Has not my soul grieved for the poor?"

–Job 30: 25

The chilly night air of Quito's high altitude cooled the interior of its homes, but the living room at HCJB's guesthouse was warm and cozy. It was the spring of 1950. Dr. Everett Fuller and his wife, Elizabeth, newly arrived American missionaries with HCJB, sat comfortably in front of the fireplace, where a fresh log blazed on the open hearth. As the couple watched the flames, a tall, lanky, young man with sandy hair cropped short for convenience, entered the room. Extending his hand, missionary pilot Nate Saint introduced himself and sat down to enjoy the fire's warmth. In the city on business, Nate was staying overnight at the guesthouse. Soon the three were engaged in animated conversation, Nate relating his experiences of jungle flying, and the Fullers, eager to learn more, interrupting him with questions.

Not one to waste time on preliminaries, Dr. Fuller questioned the young pilot about the medical needs of the people living in the jungle area where he worked. In response, Nate described some of the sick people he had seen. When he had first begun to fly the jungle routes, he had thought that Indians who needed medical treatment would take the bus to Ambato or Quito to visit the clinics there. But he had found that the jungle people would not leave their families and their familiar surroundings. Many had never traveled beyond the next hamlet; they lacked the confidence to make such a journey, warm clothing to protect themselves against the cold of the cities' high altitude, and money for treatment in a city hospital.

"The Indians are afraid of the unknown, and they won't go any farther than Shell Mera," Nate explained. "It grieves me to return the sick or injured to the jungle without their seeing a doctor, but we have no physician at Shell Mera."

"What do you do?" Ev asked.

"The missionaries care for the sick as best they can," Nate replied. "Even though many of them have no medical training, they learn by trial and error. The missionaries know that if their medicine doesn't cure the Indians, they go to the local witch doctor. But if their medicine is effective, then they can evangelize.

"When Shell Oil was in the area," Nate continued, "it operated a small clinic with a few beds. The facility was limited to company employees, and so the sick people from the jungle stayed at our house. My wife, Marj, is a nurse. Occasionally she consulted the company doctor on the treatment of a patient.

"Think of it: the government estimates that there are 65,000 people living in Ecuador's eastern region who have no doctor and no hospital! Marj and I are praying that God will supply that need."

As Nate shared his concern with the Fullers, they were deeply moved. Was God leading them, they wondered, to pioneer a medical work at the gateway to the Amazon?

The clock on the mantel struck 10. Aware that the room had cooled, the doctor rose, stirred the dying embers, placed another log on the fire, and sat down. As the flames roared, the new log snapped, sending showers of sparks up the chimney. Overlapping her cardigan sweater for warmth, Liz Fuller snuggled close beside her husband. "Go on," she said to Nate. "Tell us more."

"Well, I had some money that I inherited from my grandfather, and so, in faith, I bought 10 acres of land from the Shell Oil Company, five for Mission Aviation Fellowship to expand its base and the other five for a hospital that would serve both the community at Shell Mera and the jungle area. A hospital would support the work that the missionaries are already doing, and it would also open the door to regions where people are resisting the gospel. If HCJB will bring in the medical personnel, I'm willing to head up the construction of a clinic. We could do so much more, you know, if we had even a small hospital at Shell Mera."

Nate paused, and then asked the Fullers directly, "Would you be interested in helping with the project?"

The Fullers, inspired by Nate's account and challenged by his question, were determined to help him, but obstacles stood in the way. Dr. Fuller explained: first, he and Liz must study Spanish so that they could communicate with the people of the area. Second, he must validate his medical license in order to practice in Ecuador. Third, the mission would have to release him from Quito to work at Shell Mera. Before the evening ended, however, he promised to visit the area as soon as possible.

* * *

True to his word, a few weeks later, Dr. Fuller arranged for Dr. Paul Roberts, Ecuador's first missionary doctor, to accompany him to Shell Mera. The two had met, in 1947, at Camp Awana in Wisconsin, where HCJB's first president, Clarence Jones, discussed plans for a medical work. Although HCJB was in Ecuador primarily to proclaim the gospel by radio, Clarence Jones was burdened by the poverty and physical distress of mountain Quichuas. Indeed, to ignore their physical needs contradicted the Scriptures, for Christ had sent His disciples out "to preach the Kingdom of God and to heal the sick" (Luke 9:2). Jones had discussed the need for a doctor and a clinic with the other mission organizations, but as he explained, "In the 50 years of Protestant missions in Ecuador, no other organization had become involved in medical work." It appeared that HCJB personnel must take the initiative. And take it they did.

Dr. Roberts and his wife, Barbara, arrived in Quito in 1949, and the Fullers came the following year. Then, on April 28, 1950, HCJB began its medical work with the opening of a 16-bed Indian hostel clinic at the north end of Quito, on the cobbled Pan-American Highway. The hostel was easily accessible to mountain Quichuas as they passed by on their way to market.

Now, in June of the same year, the two doctors set out for Shell Mera to make a medical survey of the jungle area. Leaving Quito at daybreak, they traveled by car along cobblestone roads. In those days the bone-jarring trip overland was a 10- to 12-hour journey sometimes lengthened by mud slides and consequent detours. Their route took them through Pelileo, where the earthquake had occurred. Roberts, who had helped with the rescue operations, was interested to see the area again.

From Pelileo, Roberts and Fuller crossed the Patate River, drove through the town of Patate, and then followed a narrow road along the east bank of the river. Between Baños and Shell Mera, a distance of 25 miles, they crept around sharp curves on a rain-rutted road that sent up clouds of dust the moment it dried. Mountain streams gushed from rocky slopes, dotting the landscape with waterfalls. In the tropical climate, lovely lavender orchids stretched toward them, just beyond their reach. Two hours from Baños, the doctors arrived at Shell Mera. A few tiny shops lined one side of the main street. Opposite stood Shell Oil's empty administrative buildings with the landing field running behind and extending beyond the town. A quarter of a mile from the center of Shell Mera and opposite the airfield, the doctors found the Saint home. A 20-minute drive beyond Mission

Aviation Fellowship, the road ended at the frontier town of Puyo.

Nate flew the doctors from station to station in the jungle to hold clinics in the homes of missionaries. Word of their visit brought to the doors of the mission stations Indians who, until then, had avoided any contact with the foreigners. The doctors saw patients suffering from scabies and other skin problems, bronchitis and tuberculosis, anemia and machete cuts, as well as emaciated children suffering from malnutrition and parasites. The doctors listened to the lungs of persons with a hacking cough and wished for an X-ray machine to confirm their diagnoses.

In succeeding months, the two doctors alternated their visits, flying to the south jungle one month and to the north the next. The clinics they held bridged the gap between the foreigners and the Indians by reducing fear and distrust as tribal people gradually learned that the missionaries' medicines were far more effective than the witch doctors' incantations. Even though Paul Roberts was soon designing a mission hospital for the city of Quito, which would be built across the street from Radio Station HCJB, Ev Fuller continued his trips to the jungle region, becoming more and more aware of the health needs of its people.

First impressions are always lasting. On that first trip to Shell Mera, as the two doctors waited for Nate to load the airplane, they stood by the hangar and surveyed the vast expanse of jungle. Dr. Fuller could visualize the physical sufferings of thousands of people hidden in the Amazon rain forest. He pictured the isolated missionaries with their scanty supplies and limited medical knowledge. He knew they needed more accessible health care and professional advice to carry on their work among the Indians. The vision made a deep impression on the young doctor, strengthening into a God-given obsession. "Someone must help these people," he said, "and by God's help, I'll do it!"

5
Medicine Without Doctors

"Delight yourself in the Lord and he will give you the desires of your heart."

–Psalm 37:4

After the liveliness of oil company days, Shell Mera was deserted and desolate. The Shell Oil Company's empty buildings were forsaken, their windows barred shut and their doors locked. In their musty interiors spiders spun intricate webs, and hardy cockroaches, ants and other vermin ran freely through the vacant rooms. Shell Mera had become a ghost town.

Opposite the empty buildings were a few weather-beaten shops that sold *raspadura* (cakes of brown sugar), locally grown fruits, whiskey, and fresh vegetables trucked in from the mountain area. When rains made the road impassable, villagers did without these items.

On the same street, living behind storefronts, were the washerwoman with her brood of children, and the shoemaker, whose business generated little income in a place where children ran barefoot. The carpenter, too, lived on the same street, making highly polished wooden suitcases, besides running a saloon and keeping a watchful eye on his pretty daughter who was fast approaching womanhood.

The village folk lived a simple life. Their homes had no indoor plumbing; candles supplied the only light at night. No one owned a car; a bus based in Puyo provided transportation in and out of the area, traveling to Quito one day and returning the next. Mail reached Shell Mera every other week.

When Nate Saint learned that Gospel Missionary Union wanted to move its Berean Bible Institute out of Guayaquil and away from city distractions, he encouraged mission leaders to buy property from the Shell Oil Company. The mission took his advice and, in 1949, purchased the buildings nearest the road. It then assigned Keith and Doris Austin to Shell Mera to open the Bible School at its new site. Teacher Mardelle Senseney joined them in December 1950. Nate and Marj were thrilled that the Bible institute had entered the area to support the missionary cause. Shell Mera would become a Bible-teaching center instead of an oil town.

Soon Art and Hilda Schmidt joined the teaching staff. They arrived in Shell Mera for the first time on a moonless night. Since no welcoming lights showed in the dark buildings, Art asked the bus driver, "What shall we do?"

"Just call out," the driver replied.

"Keith Austin!" Art shouted.

Immediately from the pitch darkness came Keith's voice, "Here I am!"

As GMU's personnel cleaned their recently purchased buildings, Ecuador's military moved next door into Shell Oil's former administrative buildings and into company-owned houses on the other side of the airstrip. In succeeding years commercial airlines, mission-owned aircraft, and the Ecuadorian Air Force would increase activity on the landing field.

Dr. Fuller continued his visits. To assist him in his medical work, GMU provided a treatment room in the girls' dormitory of the Bible institute, with Mardelle as helper. Fuller's army experience in World War II had taught him the importance of on-the-job training. Thus, knowing that he would need a trained helper to anesthetize his surgical patients on the first day he operated, he taught Mardelle how to drip the ether. He also instructed her in postoperative care of anesthetized patients. Later, when he returned from his first furlough with a simple X-ray machine, he gave Mardelle books on radiology and assigned her the task of learning how to take X-rays. Though Mardelle learned the technique, finding a container large enough to immerse the X-ray plates in the developing fluid demanded greater ingenuity. After searching a few kitchens, she found that Doris Austin owned a dishpan big enough for the purpose. Still, there was a problem: Mardelle had to remember to clean the dishpan and return it to Doris's kitchen after each use because the large container was also needed for making bread.

Mardelle, challenged by her new assignments, learned quickly. Young and energetic, she took on a variety of duties. She taught the students at the Bible institute, supervised the girls' dormitory, ran the kitchen, took her turn at entertaining guests and cared for the sick. Although she had no formal medical training, she was eager to learn. It sometimes seemed, however, that each time a medical emergency occurred, the doctor was away.

Late one afternoon, for instance, Mardelle spotted four men carrying a young Indian girl on a homemade stretcher. Juana had suffered a gunshot wound to her right shoulder. Beneath an old rag, dirt and blood caked her skin. There was no doctor in Shell Mera, and the only car was one that George Poole and Jorge Sánchez of HCJB had driven from Quito. The two missionaries offered to take Juana to Ambato, provided that Mardelle and Juana's father would go, too.

Darkness had already settled on Shell Mera as the group drove away from the village. Juana, drifting in and out of consciousness, lay on a straw mat in the back of the car, with her father and Mardelle seated beside her. The Shell-Baños road, recently opened after repairs, was still in poor condition. As the driver swung the car around countless turns, the passengers in the rear slid from side to side like riders on a roller coaster.

Soon the patient began to vomit, and the obnoxious smell made Mardelle feel faint and dizzy. "Will this trip never end?" she asked herself desperately. Despite her best efforts at control, when Juana vomited again, so did Mardelle.

It was nearing midnight when the group arrived in Ambato. After an hour's search, they found a clinic willing to admit Juana. After this experience Mardelle and her co-workers longed all the more that Shell Mera might have a medical facility with trained personnel. But it would be a few years yet before they would receive this desire of their hearts.

Early one morning in 1953, José Vaca, a stranger to the missionaries, arrived at the door of the Bible institute, seeking help for his family. The night before, one of the foundation legs of his stilted house had collapsed, tossing a pot of scalding soup and macaroni onto his wife and children. They had had to spend a painful night as he waited for the safety of daylight to make his way to Shell Mera.

Though the town still had no doctor, Dorothy Brown, a GMU missionary and Shell Mera's first nurse, had just arrived to begin language study at the Bible institute. With her were Roger and Barbara Youderian, also new members of GMU. While Roger reassured their worried visitor, Dorothy gathered bandages, a jar of Vaseline gauze, and a few sterile instruments. When Dorothy was ready to leave, José insisted that only Roger should go. "The jungle trek will be too strenuous for you, *Señorita*" he warned. Nevertheless, Dorothy was determined to go. She had come to Ecuador to serve the Lord in medical work. Caring for the sick would be more interesting than studying Spanish, a language she found difficult.

To reach José's house, Dorothy and Roger crossed the Pastaza River in a dugout canoe hollowed from an immense log. José poled the canoe confidently, zigzagging to avoid the rapids. On the opposite shore, the three climbed a slippery embankment. Rain-drenched grass soaked the hem of Dorothy's cotton skirt, and water squished through her sneakers. Soon the two men were well ahead of her, jumping swiftly over fallen logs and brushing aside low-growing branches with such skill that they made the trek look easy. Roger, an ex-paratrooper, was nearly six feet tall. Stricken with polio at the age of nine, he walked with a slight limp. Still, he coped well with the rough jungle trail.

Lagging behind, Dorothy plodded on. Just when she thought she was making headway, her sweater caught on an overhanging branch, and her feet began to slip. She snatched for a branch to keep from falling, but now her right foot was plastered with thick gooey mud. Farther on, a battalion of leaf-cutter ants, true owners of the jungle forest, marched across her path, each carrying a half-inch section of leaf uniformly cut. She wanted to stop and watch the industrious insects, but she knew that she must not take the time.

Exhausted from trying to dodge overhanging branches, to watch where she stepped, and to keep an eye on the men, Dorothy paused, breathless. José was very nearly right: the strenuous hike *was* almost too much for her. When the men disappeared around a bend, Dorothy hurried on. At the end of the trail, in a cleared area, a cluster of thatched dwellings stood six feet above the ground on stilt foundations. A few hens scratched for food on soil already bare from their constant scrabbling. Scantily clad, barefoot children shied away from her, but one small girl with matted hair stared at Dorothy with big round eyes.

Among the dwellings, slanting sharply to one corner stood José's house. Termites, silent infiltrators into what had seemed to be a solid post, had tunneled up through the wood. The constant gnawing of these destructive white insects had felled the house. Roger and José stood looking silently at the dilapidated building.

A sturdy notched log gave access to the house next door; its lower end rested on the ground and its upper end reached through the open doorway so that it served as a stairway. The men climbed the notched post. Dorothy followed. The barest of essentials furnished the house. Along one wall was a bench: a wooden plank set on two blocks of wood. A small table stood in the center of the room. On straw mats on the floor lay the burn victims. Their next-door neighbors were caring for them.

In this primitive setting, Dorothy and Roger squatted beside the family to begin the tedious procedure of cutting open blisters to drain the fluid, cleansing with antiseptic, then applying Vaseline gauze. Roger proved to be a good assistant. In preparation for work in an isolated area, he had enrolled in a course in missionary medicine and had found that he enjoyed it. Dorothy and Roger worked non-stop. By the time they had finished, the afternoon had slipped away. As they hurried back to the Bible institute, Dorothy kept wondering about her patients' burns. Would they heal? What would a doctor have done? Without hospital facilities and an attending physician, caring for the sick was not easy!

A few days later, Dorothy and Roger returned to check their patients' progress. They were amazed to see how well the burns were healing. These new missionaries found it exciting to see how the Lord had used their efforts to restore the mother and children to good health. For several weeks they continued to visit the family.

Other emergencies interrupted Dorothy's language study. One day Marj Saint sent word that Nate was bringing in two injured Quichuas from the jungle. He would land in a few minutes. Dorothy hurried down the road to MAF. When Nate had landed and taxied to the hangar, Dorothy and Marj examined Nate's passengers, Manuel and his wife Carmela, who was eight months pregnant. The couple had been attacked and seriously wounded by a band of Aucas (Waorani Indians), a notoriously savage tribe living in the northeastern jungle. Despite two spear wounds in his chest, another in his thigh, and a hole in one hand from trying to ward off the lances, Manuel was able to walk. But a spear had lodged in Carmela's back, paralyzing her lower body. Both needed a physician's care, but no physician was available.

Nate arranged to transfer the couple to Quito, where surgeons operated on Carmela to cure her paralysis by removing the spearhead. Her wound and the ensuing surgery precipitated early labor, and several days later Carmela gave birth to a premature but healthy baby.

As Dorothy Brown and Marj Saint did their best to provide medical care for the people of the area, local women taught them a few traditional remedies. Lemon juice, used internally or externally, was said to be an effective treatment for nearly every ailment. Mud packed on a machete wound, they were told, stopped the bleeding. Babies came into the world whether or not the attending nurse was gowned and gloved according to accepted hospital procedures.

As time went by, Ev Fuller and Nate Saint's dream of a medical work on the fringes of Ecuador's rain forest did not fade. Whenever they met, the two discussed and planned the hospital's design; when they were apart, letters flew back and forth between them.

On one of Fuller's trips to the *oriente*, he took his wife and children along for a vacation. Enjoying the warm climate of the lower altitude, the children played outside all day in the hot sun. Ignoring the heat, the Fullers' eldest son, Larry, had neglected to remove the heavy underwear he usually wore in Quito. Late that afternoon, he came inside, face flushed, hands clutching his head, gasping, "Oh, my head! My head!" The lad was burning with fever: his temperature had soared to nearly 106 degrees Fahrenheit! His parents stripped off his clothes and wrapped him in a cold, wet, cotton sheet that warmed the moment it touched the lad's burning body.

Isolated from colleagues and without a medical facility, the Fullers had to care for their son alone. While the missionaries prayed, Ev and Liz spent the night pouring cold water on

Larry. As shopkeepers unlocked their squeaky doors at daybreak, and farmers shuffled down the road to their work, Larry's temperature returned to normal.

It had been an anxious night for the Fullers, one never to be forgotten. Yet the experience gave them an even keener desire to see a hospital established at the jungle's edge. Some months later, God answered their prayers. Before the Fullers left for their first furlough in the spring of 1953, the board of trustees of World Radio Missionary Fellowship, Inc. gave its approval for Dr. Fuller to concentrate on the hospital project in Shell Mera. Although the mission would now have to finance the construction of two hospitals, one in Quito and the other at the jungle's edge, its board members had the faith to believe that God was leading them into a medical ministry that He would bless. The five mission organizations working in the *oriente* assured Dr. Fuller of their full support of the Shell Mera project.

6
A Surgical "Scratch" Team

"Blessed is the man who perseveres under trial, because when he has stood the test, he will receive the crown of life that God has promised to those who love him."

–James 1:12

Work on a temporary clinic began in the summer of 1954. Under Nate Saint's direction, Indian boys with machetes slashed trees and underbrush to clear the land that Nate had bought for a hospital. Male students at the Bible institute spent mornings in the classroom; in the afternoons they joined the workers. With their flailing machetes they cut down thick shrubbery and jungle vines, among them a *mora* (blackraspberry) patch.

To help with the medical program, Dr. Ralph Eichenberger and his wife, Beth, missionaries with Wycliffe Bible Translators, moved to Shell Mera in September. Since the mission owned no housing in the area, the Ecuadorian military offered the Eichenbergers an army officer's villa on the far side of the airstrip where army personnel became their nearest neighbors. Built ten years earlier by Shell Oil, the frame house with its tile roof rested on cement posts. Though its screened windows reminded Beth of a summer cottage, she liked the house and its location. On sunny days a pleasant half-hour walk took her to the Pastaza River. By mid-December, Dr. Fuller and family had moved from Quito and settled next door.

At nine o'clock on New Year's Eve, a knock sounded at the Eichenbergers' door, interrupting Beth's letter writing. Because Ralph had recently been ill with hepatitis, Beth had planned an uneventful evening. Ralph, who still needed plenty of rest, had retired two hours earlier. Beth put down her pen and started for the door, but before she could open it, the knock sounded again.

The Eichenbergers' visitor was Dr. Fuller in an urgent dilemma. Aída Aguilar, a patient from Puyo who was seven months pregnant, had endured four days of hard labor; for the last two days she had not felt any fetal movement, nor could Dr. Fuller hear a fetal heartbeat. Yet Aída refused to go to Quito. Ev wanted to discuss the case with Ralph. Realizing the danger of Aída's condition, Beth woke her husband, and the two doctors talked. In spite of his weakness, Ralph assured Ev, "I'll assist if you decide to operate."

Since, in the circumstances, surgery seemed the only positive choice, Ev went to his home next door to alert his wife, who would be his instrument nurse. "Have you unpacked the OR instruments?" he asked, certain that the fiber drums had not yet been touched.

"No," said Liz, and a reassuring smile lit up her face. No matter how worried her husband might be, Liz's buoyancy came to the fore. Vivacious and optimistic, she was always ready for anything. Promptly she went to unpack.

On the Fullers' recent furlough, Ev had collected surgical instruments, and Liz had made surgical drapes of every size and description. Using a flashlight, she made a rapid search. Finding specific items in their unpacked baggage was far from easy, for to save space, they had mixed hospital equipment with their personal belongings. As Liz unpacked, she tossed the children's toys, blankets and clothing on the floor beside her. The family had brought enough of everything to see them through a four-year term on the field. Tonight she would need surgical gowns, gloves, sutures, drapes, basins and instruments. The 50-quart pressure cooker for sterilizing equipment was at the bottom of one barrel, but which one? She must hurry, for her husband had said that he must perform major surgery within two hours.

While Liz rummaged through the contents of their fiber drums that were stacked in the storeroom, Ev, a dynamo of energy, hurried back across the airstrip. To recruit a "scratch" surgical team, he had to assemble the missionary community. Every available person, whether professionally trained or not, would help with the operation. That night Dr. Fuller set the pattern for dealing with future emergencies. As necessary, he and his staff would call on any jungle missionary passing through Shell Mera to assist them.

Though workmen had started to build the clinic, it was not yet ready for use. The only alternative was to operate in the girls' dormitory. When Dr. Fuller arrived at the Bible institute, Dorothy Brown and Gladys Jamieson, a GMU nurse visiting from Colombia, were dragging into the laundry room a wooden desk that would serve as an instrument table.

Two days earlier Nate Saint had purposely postponed Gladys's flight to Macuma, a GMU jungle station, so that she could help Ev and Liz deliver the Saints' third child, Philip. Gladys had assisted by giving Marj the trilene anesthetic. Now waiting in Shell Mera to take the early morning bus to Quito, she was available to help with Aída's operation. "Gladys, I'll need you to give the anesthetic," Ev said.

By one o'clock in the morning, the surgical team was ready. With his photo lamp attached to a wire strung diagonally across the room, Nate Saint, dressed in a clean pair of cotton pajamas, stood ready to direct the light for the surgeon; Bob Hart was timing the pressure cooker that contained extra instruments; Dorothy, as circulating nurse, would help where needed, and Gladys was ready to start the anesthetic. Propped in a chair and still weak from his bout with hepatitis, Ralph Eichenberger sat opposite Ev, ready to assist. Liz had arranged the sterile instruments neatly on the desk. Turning to his wife, Dr. Fuller checked to make certain that she had the suture he needed. Then he looked around the laundry room with its makeshift equipment and a surgical team recruited from a scattering of expatriates. What a contrast to a modern hospital in the homeland!

Ev Fuller was stirred by the enormity of this undertaking. This would be the first major operation performed in the area, and it would leave an indelible and lifelong impression on him. He believed firmly that the Lord had called him to serve these people, and with God's help, he could and would do so. As he picked up a scalpel to make the incision, he looked around at the assembled team and said, "Let's pray." The prayer he spoke was brief, a simple request for guidance. As the scalpel plunged through the patient's skin and fatty tissue, he asked what time it was. Exactly two hours had elapsed since he had decided to operate.

During surgery Aída Aguilar delivered a premature and abnormal infant; the baby was stillborn. The team finished operating at 3:00 a.m., and all its members, except Gladys, were in bed by 4:15 a.m. Gladys had left the makeshift operating room to catch the bus for Quito at daybreak; she would not see her bed for several hours. Aída, who suffered no postoperative complications from that eventful New Year's morning, returned home a few days later.

Inspired by the success of their first major surgery, the doctors worked hard to finish the clinic. While Nate continued to oversee the project, Jim Elliot, a Brethren missionary, came out from his jungle station to install the new building's electrical wiring. Missionaries serving with the Christian and Missionary Alliance in Tena and Dos Ríos constructed the doors and windows. The Gospel Missionary Union's staff at the Bible institute gave their time and effort. When Nate returned from his afternoon flights, Marj prepared cool lemonade for him to take to the construction workers perspiring in the tropical heat.

Every week the doctors expected to open the new facility, as the construction took top priority. They were seeing more and more patients and doing operations in the girls' dormitory. A number of smallpox cases in the area required an immunization program, but the morning Dr. Fuller planned to vaccinate the local children, several minor emergencies interrupted his schedule. *Señora* Vásquez rushed in carrying her two-year-old son, Benito, whose foot had been run over by the mail bus; X-rays showed no broken bones. Then Mike, a missionary child, fell, cutting his lip. Ev Fuller sutured the cut, but when Mike returned to his play, he bit out all the stitches. Ev sutured the lip again, this time tying tighter knots.

Concerned about the outbreak of smallpox, the Dominican priest at Mera asked Dr. Fuller to vaccinate the Roman Catholic school children there. Since he was attending patients in the clinic, Ev sent Liz to the village, 30 minutes up the road. That morning she vaccinated 96 children and a few adults.

During the first two weeks of March, Ev Fuller and Ralph Eichenberger helped Nate by working half time on the construction of the new clinic and half time providing medical care. By the middle of March 1955, the clinic was set up and ready for use. The two-room building held a doctor's examining room partitioned with a curtain to provide a small laboratory as well. A ladder beneath a trap door gave access to a poorly ventilated second floor that served as a staff dressing room and storage space. The other room was reserved for operations and deliveries.

Ingenious Nate constructed an operating and delivery table of welded angle iron. Like a gynecologist's examining table, it could be lowered at one end for obstetrical cases. When it was needed for surgery, the nurse fastened the drop section underneath the table. Still there was a problem: if a patient became too restless, the table disengaged, and its lower section collapsed, leaving the surgeon jittery and the partially anesthetized patient wondering what had happened. More than once a staff member sat under the table to keep it elevated during an operation.

Dorothy drew the assignment of sitting under that table while the doctors operated on Wantico, a Jívaro witch doctor. As Wantico went under the anesthetic, he threshed from side to side, and Dorothy had to struggle hard to keep the table from collapsing. When Wantico awoke after his surgery, he described his experience to Dorothy. "I thought I was in my dugout canoe," he said, "fighting the rapids on the Upano River. I was sure I was going to drown!"

On opening day, 60 patients came to the new clinic for help. On two succeeding nights, the doctors and Nate had worked until 2:00 a.m., pouring the cement floor in the room designated for surgery and deliveries. The cement had barely hardened when washerwoman Victoria Saramiento arrived with her little daughter Mercedes. Earlier in the day Mercedes had eaten something toxic that disturbed her intestinal worms, causing them to head for the exit. So eager were the parasites to leave that they piled up in Mercedes' abdomen, blocking her intestines and necessitating surgery.

Following the operation, staff placed Mercedes on a portable stretcher. It took some maneuvering to hold an intravenous bottle and keep the child level as they carried the stretcher out the clinic's front door and down the steps, and placed it in the back of the pickup truck that would drive her through the village to the Esmeraldita ("Little Emerald") Hotel.

The Esmeraldita, owned by the Quillones family, was Shell Mera's wayside inn, named after the proprietor's hometown of Esmeraldas ("emeralds"). Dr. Fuller had arranged with *Señor* Quillones to use the ground-level front room to accommodate postoperative patients. Patients and travelers slept together in one large open area partitioned only by a suspended sheet to provide separate quarters for men and women. The patient's IV bottle hung on a supporting post. As the number of surgical cases increased, Dr. Fuller would sometimes have four or five postoperative patients at one time, all of them being cared for by family members. If an operation was extensive and the patient needed special observation, a nurse stayed at the hotel to give the necessary care.

While the "whites," as the *mestizos* were called, used the hotel, the Indians occupied native-style bamboo huts called *casitas* that the workmen had built at the edge of the wooded ravine. These *casitas* made jungle patients and their families feel at home during an extended convalescence. Although these buildings achieved their purpose, the temporary residents often failed to care for them. When they lacked firewood to build a fire in the center of the mud floor, for instance, they simply tore the bamboo siding off the walls and burned it.

Before the doctors started their morning clinic, they visited their patients in the *casitas* and those at the "Little Emerald" Hotel. This routine persisted until the hospital was completed three years later.

The clinic's new operating room attracted a good deal of interest. The Fuller children and their friends, curious about surgeries taking place there, would climb up an old bamboo ladder and peek through a screened window. The child on the top rung described to the others what was happening. When he felt faint, he climbed down, and the child on the rung below replaced him.

Unaware of this audience, Dr. Fuller was performing an exploratory laparotomy one day when one of the boys impulsively called out, "What did you find, Dad?"

The doctor's reply was honest. "Nothing," he said.

There was a moment of silence before the next question came. "Well, Dad, why did you operate?"

Another time, in the middle of an operation, the Fullers heard a young voice at the window say, "Move over, Dad. I can't see any blood!"

* * *

Workmen began the construction of the hospital on April 5, 1955. On that day they poured a foundation that consisted of several cement posts to elevate the building a few feet above ground level. Grooves around the perimeter of the posts would be filled with oil to keep termites out. (Years later the cement posts corroded, and termites invaded the wooden structure.) When the cement had hardened, Nate and his willing but largely unskilled helpers laid joists on the supporting posts and covered them with floor boards. The work took longer than expected because of wet weather.

Roger Youderian, the best qualified person in construction, did most of the framing for the hospital. He and his wife, Barbara, had been working at Macuma among the unreached Jívaros. Leaving Macuma in the fall of 1955, Roger flew to Shell Mera. Dressed in a T-shirt and blue jeans, he worked alone day after day, securely nailing sheets of corrugated aluminum roofing. A man of unbelievable dedication, Roger was intent on finishing the roofing as soon as possible, for he knew that once the roof was in place, workers could finish the interior whether it was raining or not.

Roger hoped to give even more time to the hospital project in the following year, but 1956 was to bring many changes. Roger would not be there to help build a hospital on the fringe of the jungle; Nate Saint would not see the fulfillment of his dream of adequate health care for the isolated people of the rain forest. Work on the new building would come to an abrupt halt.

7
A Promise Fulfilled

"Unless a kernel of wheat falls to the ground and dies, it remains only a single seed. But if it dies, it produces many seeds."

–John 12:24

Dorothy Brown trudged down the road toward her room at the Bible institute. With each step, dust swirled about her feet, powdering her blue sneakers a smudgy gray. Soon the rainy season would begin, but right now she wished for a gentle shower to settle the dust of the road.

Dorothy spent Sunday afternoons teaching a children's Bible class in the Balseca home at Moravia, a mile up the road from Shell Mera. While she taught the children, Nate Saint usually read and explained the Scriptures to the adults, but on this Sunday, January 8, 1956, he had not appeared. It was unlike Nate to miss a class, for he considered teaching the Word of God just as important as flying for the mission. Walking home alone after the class, Dorothy asked herself, "What could have happened to Nate?...Could he have been called out on an emergency flight?"

Meanwhile, down the road at the Saint home, Marj eagerly switched on the radio receiver for her prearranged 4:30 p.m. contact with Nate. Missionary colleague Olive Fleming, whose husband, Pete, was in the jungle with Nate, stood beside Marj. Both women were eager to hear the latest news from their spouses.

They heard nothing. Thinking that she might be off frequency, Marj tuned the set for better reception, but still there was no sound. The women waited. Five minutes passed, and then another five. Olive suggested that perhaps Nate's watch was slow, or that he might be having difficulty with his radio. Marj called into the microphone, "Shell Mera calling and tuning for 56 Henry." There was no answer--only silence.

Aware of the dangers of flying across thick jungle forests and winding streams, and of facing unexpected storms, Nate always kept in touch with Marj by radio, informing her of his location and activities. In his seven years of jungle flying, he had never failed to contact his wife as arranged. Gradually the two women's eagerness turned to anxiety. To reassure themselves, Marj and Olive thought of all the possible reasons why Nate might have been unable to make radio contact, and the reasons they suggested to each other helped to calm their fears.

Just after sundown, Dr. Arthur Johnston of HCJB walked into the radio room. Seeking peace and quiet from the usual bustle of city life, he had come from Quito to Shell Mera for some uninterrupted hours of study in preparation for the medical exams he must write to qualify for an Ecuadorian license to practice. Aware of Marj's usual friendly outgoing personality, he was surprised to find her sitting with her head down on the desk and the silent radio turned on. Concerned, he gently asked, "Is something wrong, Marj?"

"I don't know," Marj replied. "Nate's in the jungle with Ed McCully, Jim Elliot, Pete Fleming and Roger Youderian. They went together to try to make contact with the Aucas. I expected Nate to call me this afternoon, but I haven't heard from him."

The Aucas, a Stone Age people, were the most feared of the Amazonian tribes. Since 1542, when the adventurous Francisco Orellana traveled down the Napo River to the mighty Amazon, the Aucas had taken an aggressive stance against all outsiders, consistently refusing, for four centuries, any proffered friendship. To defend their territory, they had speared the conquistadores, their Quichua neighbors, and modern prospectors searching for rubber, oil and gold.

Years earlier, when Reuben and Grace Larson settled in the area, they had built their home on the opposite side of the river from the fierce unknown tribe. One day when the Larsons returned from a trip, they found spears stashed nearby and close to them, the biggest footprints they had ever seen. They traced on paper the outline of a broad foot with splayed toes and sent it to their friends, asking them to pray that God would reach the Aucas with His message of love.

Now, in January 1956, a friendly contact seemed a likely possibility. For several months, Nate, Jim and Ed had been making their preparations. Nate had invented a "bucket-drop" device that consisted of a basket on a spiraling line with a mechanism that automatically released its contents the moment it touched ground. He had first used it in Achuar territory and then, to prepare for a series of gift drops to the Aucas, he had practiced for days at the Shell Mera airstrip. With the bucket tied securely on one end of the cord, Nate flew in a wide circle while his helper reeled out the line from the door of the airplane. As the bucket dropped earthward, Nate closed the circle, and the basket came to rest on the runway.

During the 13 weeks after the men had perfected their delivery technique, Nate and Ed McCully made weekly gift drops to the Aucas. In return, the Indians sent back such items as a live parrot, two bananas for the bird, smoked monkey meat, feather crowns and clay pottery. With this exchange of gifts, the men were convinced that it was God's time for them to enter Aucaland. To avoid interference from outsiders, the five men kept their mission secret from the public and even from missionary colleagues. In his radio contacts with Marj, Nate referred to their location as Palm Beach.

Eleven months before "Operation Auca," in February 1955, Nate's sister, Rachel Saint, began learning the Auca language from Dayuma, an Auca girl who had escaped from the tribe 12 years earlier, when her family became the target of vicious spearings. As the men planned their contact, they considered taking Rachel into their confidence. She knew more Auca words and phrases than any person outside the tribe, but in the end, they decided against involving her.

On Tuesday, January 3, 1956 Nate successfully landed his tiny aircraft on a sandy shore along the Curaray River. The sandbar was high, dry and level. Four days later three Aucas (two women and a man) came out of the jungle and met the missionaries. The Aucas seemed friendly. The missionaries expected that the Indians would return that Sunday afternoon.

Now, on Sunday evening, as Marj related this information to Art Johnston, she asked him not to tell anyone. Outwardly she remained composed, but inwardly she was worried. She assured Art that she would be alright, but she slept very little that night.

By 7 o'clock on Monday morning, January 9, Nate's colleague Johnny Keenan was piloting his plane toward "Palm Beach." Johnny had come to Ecuador six months earlier to help Nate with the mission's growing number of flights. Nate had shared with Johnny the plans to contact the Aucas and had pointed out the sandy strip, 60 air miles northeast of Shell Mera.

Flying in that direction, Johnny searched for the beach and soon located it. From the air he spotted Nate's bright yellow Piper plane on the sandbar. Circling, he flew lower and saw that the aircraft lay damaged, stripped of its fabric; there was no sign of the men. With a heavy heart, Johnny picked up the microphone and called Marj at Shell Mera to tell her the news.

A search party was quickly formed. The ground party, led by GMU missionary Frank Drown of Macuma, was made up of seven missionaries, including Dr. Art Johnston, 13 Ecuadorian soldiers and several Quichuas. For three days the men struggled on foot through sodden forest and downriver by dugout canoes. Gripped by fear of an Auca attack, they kept a constant vigil. When the helicopter of the United States Rescue Squadron, hovering overhead, reported seeing speared bodies in the river, their worst fears were confirmed. They lost all hope of finding any of the men alive. On reaching "Palm Beach," the ground party located four bodies. The fifth, that of Ed McCully, had been identified earlier by a group of Indians from Arajuno. It had been washed downstream by the rain and rising river. As the missionaries buried their colleagues in a common grave, a sudden tropical storm of violent wind and rain darkened the sky. In an atmosphere of uneasiness and fear, Frank Drown recited from memory a brief burial service.

"Operation Auca" had been a carefully conceived mission with every detail planned from the moment Nate's airplane touched down until the men had removed the tall trees at the end of the beach and completed a tree house. Why had Indians who appeared friendly on Friday suddenly turned on the outsiders and speared them on Sunday? The question would remain a puzzle for several years.

In all of its history, Ecuador had never made the front pages of the world's newspapers. Now news from Ecuador shook the world: five Americans between the ages of 27 and 32 had died in the jungle when hardwood lances approximately nine feet long pierced their bodies. Reports of the men's courageous attempt to communicate God's love to a tribe of Indians produced varied responses. Why should those anxious to carry the gospel to an unreached tribe die so early in their careers? To some it seemed a terrible waste of missionary manpower: at that time Jim, Ed and Pete were the only male missionaries in Ecuador who spoke the Quichua dialect; Roger was in contact with the head-shrinking Jívaro and Achuar tribes; and Mission Aviation Fellowship had lost one of its finest pilots. Was this event a waste of missionary manpower? Were the labors of these men in vain? Would the grains of wheat that had fallen into the ground both materially and spiritually bring forth much fruit? Despite temporary loss, in God's planning nothing is ever wasted.

The Aucas had been a hazard to early explorers, and failure to subdue them had embarrassed their countrymen. Yet to the five valiant men who planned and conducted "Operation Auca," their attempt to bridge the gap between the outside world and the Stone Age tribe was a challenge. For world missions the impact was immeasurable. As news of the terrible incident spread, men and women rededicated themselves to world evangelism. With a renewed interest in mission work, countless young people now volunteered to replace the men who had given their lives to spread the gospel.

When news of the tragedy spread throughout the Shell Mera community, the first reaction was numbness. The five widows appeared to cope with their grief better than their colleagues, who easily recalled some of their last unforgettable moments with their friends who had died.

Mr. Senseney, Mardelle's father, had come to Ecuador to help build the new hospital. For Nate's last trip, he had helped to load the aircraft with food, radios, parts for a prefabricated tree house, clothing, gifts for the Indians, utensils, medical supplies and a secondhand movie camera that wouldn't film at the speed required to synchronize with sound. He had been puzzled at some of the equipment Nate was taking into the jungle, but had said nothing.

Mardelle remembered how sad Nate was when he had failed to resuscitate a newborn twin at the clinic. In fact, Nate felt worse than the parents, who were happy to have only one baby to take home to their already large family.

The unexpected death of the five men made an impact on the students at the Bible institute. In disbelief they said, "Oh, how could the Lord let those men die!" Of all the students, Raúl Del Salto was perhaps the most deeply affected. Feeling that

God was calling him to replace one of the men, he went to Puyupungu where Pete and Olive Fleming had worked.

A few weeks before the tragedy occurred, Dorothy Brown had asked Nate to talk to Sergio Villalba, a carpenter's son, who was suffering from osteomyelitis (a severe bone infection) in his right leg. While Sergio was staying with his parents in the *casitas*, he screamed so loudly with pain that he could be heard at the Bible institute. Nate came and talked with the boy.

"You know, Sergio," Nate said, "I suffered with the same sickness you have when I was young like you. I told the Lord that if He would raise me up, I would serve Him as long as I lived. Indeed, I promised Him that I would go to an unreached tribe, and I still have to do that."

The carpenter remembered Nate's words. "Surely Nate has fulfilled his promise to God," he said.

In the aftermath of "Operation Auca," Mission Aviation Fellowship assigned Hobey Lowrance to take Nate's place in Ecuador. After an absence of several years, the Lowrances returned to Shell Mera, where Hobey continued to make aerial gift drops to the Aucas. These gift drops and the prayers of hundreds of people would eventually bring results.

In November 1957, nearly two years after "Operation Auca," two native women walked out of the tribe. Had Rachel Saint or Dayuma been in the area, they could have communicated with the women, but they had gone to the United States to appear on the TV program "This Is Your Life." During an influenza epidemic, Dayuma became ill; for weeks she hovered between life and death. The month in the States became a year.

Jim Elliot's widow, Elisabeth, who has documented "Operation Auca" in her book *Through Gates of Splendor*, was at a nearby jungle station. She went immediately to welcome the visitors. Though she could not understand them, she rejoiced at the opportunity for contact with the tribe that had killed her husband. She began at once to learn Waorani (Auca), language.

Later the Indians invited Rachel, Elisabeth and Dayuma to live with the tribe. As the women gained a knowledge of the language, details of the massacre unfolded. Intertribal warfare had depleted the male population of this tribe, one of four groups of Aucas, so that of its 60 members, 53 were women or children. The friendship on the beach that fateful Friday in January 1956 had been genuine until one of the missionaries pulled a picture of Dayuma from his pocket. Unknown to the Americans, the older woman on the beach was Dayuma's aunt, and the younger one was her little sister. The Aucas, who wore only a G-string tied around their waist, thought that the picture had come from the missionary's body. They concluded that these men had eaten Dayuma. They returned to the tribe that night and reported that the foreigners were cannibals. Two days later, the massacre occurred.

When the Rescue Squadron, reporters and visitors left Shell Mera, life slowly returned to normal, except that now the missionary team faced an unfinished hospital. The skeletal structure stood as a silent reminder of the sacrifice the men had made. Nate had envisioned a medical service based at the entrance to the jungle; his enthusiasm had inspired the team. Now uppermost in the minds of the Shell Mera personnel were two questions: How could they finish the hospital? Who would supervise its construction?

8
The Water Tower

"Let your light shine before men, that they may see your good deeds and praise your Father in heaven."
–Matthew 5:16

After the martyrdom of the five Americans, missionaries and Bible institute students worked on the Shell Mera hospital. But the project, lacking leadership, moved slowly. By comparison, the medical work in Quito had moved forward. Under the direction of Dr. Paul Roberts, Rimmer Memorial, Ecuador's first evangelical hospital, had opened its doors to the public on October 12, 1955. Earlier, Shell Mera personnel had appealed to HCJB President Clarence Jones for help in building the jungle hospital. Quito staff, however, were extending themselves to the limit. Even Clarence Jones and his wife, Katherine, immersed themselves in the operation of the new facility. Katherine took charge of the kitchen and laundry, while her husband, with the aid of a few textbooks, ran the business and maintenance, thus freeing Paul Roberts for medical duties. To Shell Mera's request for more help, Clarence Jones replied, "We don't know how to run *this* place yet!"

For the time being, therefore, the staff at Shell Mera would have to manage on their own. Still, the mission did assign a new doctor-nurse team to the clinic there. In February 1956, after two years in Quito, Dr. Art Johnston and his wife, Verna, moved to the jungle's edge to live, like the Fullers, in military housing across the airstrip from the clinic.

As the clinic staff struggled to finish work on the hospital, friends from Quito came to help. Jack Shalanko of HCJB's Russian Language Service was one of the helpers. On his first visit, Jack, who had a keen eye for perpendicular lines, walked to one end of the long corridor and looked at the hospital structure. At this point, no supports had been installed, and the rooms were not partitioned. Because the end walls had not been properly braced, the building tilted diagonally. Suddenly Jack's voice boomed the length of the corridor, "This hospital is collapsing! It's going to fall over!"

To prove his point, Jack dropped a plumb line from the mid-point of the corridor ceiling. To correct the problem he had the workmen attach block and tackle from the center of the second floor to a tree at the rear, hook the pick-up truck to the ropes, and pull the tilted walls into place. Then they installed braces to keep the building upright.

Meanwhile mission leaders made known the need for construction workers. Well-qualified and experienced men volunteered their services, but since they did not know Spanish, these working visitors relied on the busy missionaries to buy the building materials needed and to translate to and from Spanish when they had to communicate with the Ecuadorian workers. One contractor was easily riled by these workers, and his inability to communicate in Spanish, combined with his impatience with the Ecuadorians and their distinctly different culture, created some thorny situations.

Despite such difficulties, friends and visitors did make a worthwhile contribution to the hospital's completion. One volunteer, 80-year-old Jim Marshall, traveled from the United States to the edge of the rain forest to install the electrical wiring in the new building.

To speed the progress of the work, HCJB directors wisely appointed Bob Wittig to supervise construction of the Shell Mera hospital. The *oriente* was familiar territory to Bob and Keitha Wittig who had replaced the Saints during their furlough in 1952. Although Bob, who was a pilot, lacked a gift for linguistics, he loved the Ecuadorian people, especially the Indian boys who worked with him. They, in turn, responded to his leadership with full cooperation.

When a driver was needed to haul stones, gravel and sand from the river to the construction site, Liz Fuller volunteered, alternating with Keitha Wittig. Even the two doctors integrated volunteer construction duties with their practice of medicine. When construction materials were needed, Ev Fuller and Art Johnston traveled out of the area to buy them.

One dark, rainy night when the two doctors were on their way to Shell Mera, Ev Fuller stopped his car at a ford on the Río Blanco (White River) before driving through the water. Torrential downpours had brought tons of rock tumbling down the mountainside, and jagged stones jutted from the riverbed. Since the road was deserted, Art Johnston valiantly took off his trousers and plunged into the river. Working in the glare of the car's headlights, he had begun to move a few heavy rocks out of the car's path when a local bus came round the bend, pulled in behind Ev's car, and shone its lights full on him. The modest young doctor was embarrassed to be seen in his underwear by the local people, some of whom were his patients.

Yet the biggest challenge was not the rocks in the river. A few miles below Baños is Agoyán Falls. The waterfall, a favorite tourist attraction, changed shape every year as a result of constant erosion. At this site, on June 13, 1953, after eight days of continuous rain, a large chunk of the mountain fell away. In the avalanche, huge boulders and the rising river swept away a bridge, the road and all obstructions in its path. Since it was impossible to move this mound of solid rock, workmen laboriously tunneled through it. With pick and shovel they chipped out a passage approximately a thousand feet long. The work took between three and four years to complete. During that period, Agoyán's frothy waterfall moved upstream from its earlier location east of the tunnel to the west side. (Years later, when the new Shell hospital was under construction, experienced engineers tunneled through this same rock to clear a second tunnel for traffic and a third one for a hydroelectric plant that was also being built at that time.)

After the landslide, local people and missionaries needing to pass the falls had to climb down steep cliffs, creep across improvised footbridges on slippery logs straddling deep chasms and struggle up huge mounds of fallen earth, often through mud and rain. The journey in and out of Shell Mera exhausted the best of travelers.

To provide an alternative to climbing across the landslide, the local people installed a crude, handmade *tarabita* (cable car). Its wooden platform had no protecting side rails. The passenger hung onto a bar with one hand and with the other pulled a strong twisted rope through pulleys to move the *tarabita* forward. The crossing was a frightening experience, especially at the moment the platform swung out over the huge abyss where, hundreds of feet below, the turbulent Pastaza River roared. Using this precarious method, the Shell Mera staff brought construction materials and supplies to the mission. Travelers rode the local bus to Agoyán Falls and then either crossed on the cable car or made a *transbordo.*

Loaded with bulky parcels, they and their carriers zigzagged up and across the 400-foot mound of rock and earth, and buses on both sides of the obstruction picked up stranded passengers and switched loads.

While construction of the hospital was still in progress, Dr. Fuller learned that an American road-building company had finished its contract near the coast and was about to auction off much of its equipment. Since this included a 12,000-gallon water tank, an item unobtainable in Ecuador, he and Art Johnston decided to attend the sale. To reach the auction, the two doctors took a bus to Agoyán Falls, made a *transbordo*, picked up Ev Fuller's car at Baños and drove to the coast, a two-day journey. When they arrived, the sale was already in progress. The doctors waited all morning for the water tank to be auctioned. When noon came and it was siesta time for the coastal people, they drove down the road to look for a place to eat.

On their way, the two doctors stopped to help a man whose car was stuck in the mud. His name, he told them, was Oscar Terán. Later Oscar joined Ev and Art in the restaurant. "Why are you here?" he asked.

"We're here for the auction," Ev Fuller answered. "We hope to buy the water tank. Why are you here?"

"For the same reason you are," Oscar replied. "I want to buy that tank, too. It's a mighty big one! What do you want it for?"

When the doctors explained that they wanted the tank for the new mission hospital that was being built at Shell Mera, Oscar responded warmly. "You people do a good work," he said. "The hospital will help my people. I'll not bid against you." When Ev and Art went back to the auction, they saw that Oscar, true to his word, had not returned.

"Do you think that we'll be able to get the tank for $500?" Art whispered to Ev. It was all the money the hospital funds could manage. The auctioneer mounted his stand, and the bidding began. For a few suspenseful minutes the two doctors silently prayed as they made their bids. Suddenly the bidding stopped, and the auctioneer nodded in their direction. To their surprise and delight, the water tank was theirs--at a price of exactly $500! The Lord had answered their prayers. The Shell Mera hospital would have its own water supply.

Now began the task of dismantling the huge tub, nut by nut and bolt by bolt. When the task was done, they loaded the sheets of steel onto a hired truck for the long trip across the

Andes Mountains down to the jungle's edge. Ev Fuller rode with the truck driver; Art Johnston drove Ev's car. The truck moved slowly, and soon Art was far ahead. When he reached the *páramo*, a bleak plateau, he pulled to the side of the road to wait for the truck. At an altitude of 12,000 feet, the frigid air was a drastic change from the hot, muggy atmosphere of the coast. Dressed for a warm climate, Art sat shivering in the dark. He had not planned to spend the night on this windswept mountain, but he wanted to be there when the truck came by. Then he remembered that there were newspapers in the car. Though they were wet, he wrapped himself in them. Good insulators against the cold, the newspapers warmed him, and he slept.

Meanwhile, all night long, the truck lumbered slowly up the mountain in low gear. At daybreak it reached the top of the *páramo* and shuddered to a stop beside the parked car. By now both doctors and the truck driver were famished. They drove down the mountain to a wayfarer's inn at a fork in the road, a stopping place for weary travelers in search of a night's lodging.

In the dull and gloomy interior, with its cold damp floors, they asked the cook, "What's for breakfast, *Señora*?" She pointed to an enormous pot where two chickens, chunks of corn cobs, and potatoes rolled in boiling liquid. Dipping her ladle into the pot, she filled three bowls. To the hungry travelers on that cold morning, breakfast had never tasted so good!

After breakfast, the men, their hunger now satisfied, continued their trip, reaching Baños by mid-afternoon. Leaving Ev's car at the Gertrudes Hotel in Baños, both doctors traveled in the truck to Agoyán Falls where the road ended abruptly at the landslide. There they off-loaded the disassembled tank so that the truck driver could return to the coast. Then the doctors began the tedious task of arranging the sheets of steel and loading them, a few at a time, onto the *tarabita*. When the first load was ready, Ev Fuller looked at the depth of the gorge below. There was no way he was going to risk his life on that *tarabita*! Turning to Art, he said, "You know, I have children."

"You have children, and so have I," Art retorted, but he didn't argue. Instead, he stepped onto the platform and rode across the gorge with the first load. The Agoyán had two cable cars. One car took Art and the cargo past the face of the cliff to a little shelf where he transferred the load onto another platform that carried it to the far side of the avalanche. It took many trips back and forth before Art had all the steel on the Shell Mera side. When the doctors finished transferring their load, it was late in the day. Ev climbed across the landslide while Art rode around it on the *tarabita*. Since no trucks would now arrive before the next day, Art crawled inside a roll of reed mats for the night, and Ev slept inside the partly excavated tunnel.

Early the next morning, when the trucks lumbered up the road from the *oriente*, the doctors hired one to transport their cargo of steel to Shell Mera. Once the disassembled water tank was on hospital property, workmen had the laborious task of reassembling it. Next they needed to build a tower to support the tank, but they had no suitable materials. On flights across Arajuno, a village in the north jungle, Bob Wittig had sighted, at the side of the landing strip, piles of steel left from Shell Oil Company days. He arranged for Gonzalo Ruales, an Ecuadorian pilot, founder and owner of *Transportes Aéreos Orientales* (Eastern Air Transport)--TAO for short--to bring the steel from Arajuno to Shell Mera in his Junker aircraft.

The Junker was ideal for transporting bulky cargo. Used by the Germans to transport paratroopers during World War II, it was a shell of an airplane with corrugated aluminum siding, no interior lining, and only side benches to accommodate passengers. With this plane Gonzalo transported the discarded steel from Arajuno, and the workmen used it to build the water tower. When the tower was complete and the tank in place, water from a cistern was pumped into it. The height of the tower created excellent pressure in the hospital's water taps.

In a tropical climate where the annual rainfall is approximately 260 inches, it is easy to forget that droughts sometimes occur, requiring workmen to haul water from the river in metal drums. To provide a constant water supply, the hospital's maintenance crew helped the townspeople, in 1972, to pipe water from the mountains. Their good deed provided the town with the luxury of running water in their homes for the first time and gave the hospital a back-up system.

The water tower not only provided water for the hospital and the surrounding mission-owned houses, but also afforded anyone brave enough to climb its steel structure a panoramic view from its top. Later, a floodlight was connected to the tower. Like a light shining in a dark place, it became a beacon for anyone entering the town in search of medical care. In its spiritual ministry, too, the hospital pointed men and women to the One who is the Light of the world.

9
Launching Epp Memorial Hospital

"Launch out into the deep, and let down your nets for a draught."

–Luke 5:4b, KJV

The second anniversary of the death of the five American missionaries came and went, and still the hospital at Shell Mera was not finished. The workers had no way of obtaining fine-finished lumber. The only wood available was rough-planed boards supplied by a Czechoslovakian sawmill owner in the nearby hamlet of Moravia. Finishing this lumber was a slow process; even before they began to cut and plane, the men checked the wood carefully, choosing boards cut from the center of the tree where the wood is hard and free of termites. Interior paneling was not obtainable; the exterior walls were made of single boards nailed across the supporting studs. Through cracks between these boards, cold air seeped into the building on clear, cool, moonlit nights and cloudy damp days.

To protect the new wood from mold and mildew, the women spent hours varnishing, inside and outside, the rough lumber of which the rustic frame building was made. Then they cleaned and polished the windows until they shone. No sooner had they finished this task, however, than Dave Findlay, a newly arrived missionary from Canada--his attention caught by something outside and failing to see the clear glass in the window frame--stuck his head through a sparkling pane and shattered it.

There were other frustrations, too. For one, the corridor linoleum, on order from Scotland, was taking much longer to arrive than anyone had expected. Still, trusting that the firm would eventually supply the material ordered, workmen partitioned off the rooms so that the width of the corridor corresponded to the measurement of the linoleum on order. When the linoleum finally arrived, months after the hospital was officially opened, it came in one piece. The men rolled it along the corridor from the main entrance to the operating room, and to everyone's relief, it fit perfectly. A lack of funds, however, meant that staff would not enjoy the luxury of linoleum on the wooden floors in patients' rooms.

During the hospital's planning stage, Rev. G. Christian Weiss of "Back to the Bible Broadcast" visited the *Oriente*. While in Shell Mera, he talked with Dr. Fuller and became interested in the hospital project. Back in the States, Rev. Weiss encouraged the missions committee of "Back to the Bible" to help finance the hospital's construction. The organization's first donation amounted to $10,000, which was used to pay for the exterior lumber; it took an additional $25,000 to finish the interior. This amount arrived during a period of months in small donations from friends of the missionaries, HCJB supporters and additional gifts from "Back to the Bible" listeners.

Mission leaders invited Dr. Theodore Epp, director of "Back to the Bible," to be the guest speaker at the hospital's dedication, set for May 10, 1958. Dr. Epp and his wife arrived a week early to visit in the area. The day before the ceremony, colleagues from Quito trudged across the landslide at Agoyán Falls, carrying by hand delicacies that the women planned to serve their guests. MAF pilots were flying several officials from Ecuador's capital and, weather permitting, would take them back later the same day.

On the morning of the new hospital's official opening, staff set up two wooden cots in the men's ward and brought two young patients from the *casitas* to occupy the beds while the guests toured the building. The opening ceremony began at 1:30 p.m., three hours later than planned. In Ecuador no function begins until all the participants have arrived; no one is ever said to be late, and in Shell Mera the time of guests' arrival depends largely on the weather.

In his message Dr. Epp said that his father, Jacob Epp, missionary to the Hopi Indians, spent the last few years of his life as prayer leader at "Back to the Bible." He devoted many hours each day to prayer in support of requests from radio listeners. In memory of Mr. Epp, and in appreciation of the many donor gifts received through "Back to the Bible Broadcast," HCJB named its new health care center Epp Memorial Hospital. A simple plaque was placed at the entrance, bearing the Spanish name *Hospital Vozandes del Oriente* and, below the name, the words "To the glory of God and to the service of the people of Ecuador. In memory of J.B. Epp."

About sixty people gathered at the hospital's front entrance for the opening ceremony. Few were local people, even though this medical facility had been built to provide them and their families with health care. The number of local people was small because some were suspicious of evangelicals; others felt out of place attending a function along with persons of higher social status; and still others were wary of modern medicine.

Twenty-five years later, at the hospital's anniversary celebration, Zulay Cárdenas, president of Epp Memorial Hospital's Employees' Association, described those early years in these words:

> It was only by the mercy of God that the small population of this area was not wiped out completely. Methods of curing people were very limited. There were only two facilities where local people could obtain medicines--the Red Cross in Puyo and, occasionally, from the pharmacy at the military hospital. For the people of the *Oriente* to

> have a hospital built at this location was like a gift dropped from heaven.

To the missionaries present on the day of its official opening, the new hospital was a facility beyond their highest hopes.

Three weeks passed before staff had the hospital suitably equipped to admit patients. Meanwhile, the doctors continued to deliver babies and to operate in the clinic. Until the hospital had enough beds, staff continued to take postoperative patients to the Esmeraldita, a nearby hotel, for recuperation.

Help in putting the new facility in order came from most unexpected persons. The first of these was Bill, a young transient from Rochester, New York, who was traveling through South America to see "what made the world tick" and how people lived in faraway places. When Bill stopped and knocked at the front door of the Fuller house, the family welcomed him warmly. Before long he was helping move equipment and doing odd jobs. The experience was a turning point in Bill's life: he was searching for something of value, and he saw that these missionaries had found it. Before he left Shell Mera, Bill had put his trust in Christ.

Another helper, Al Dickson, a Christian medical student, wanted to visit a foreign mission field. His arrival in Shell Mera coincided with the transfer of the operating room equipment from the clinic to the hospital. He helped the staff move everything except Nate Saint's handmade collapsible table, which was left in the clinic, where the doctors would continue to use it in their work with outpatients. Though Al Dickson's visit was short, his enthusiasm for jungle medical work did not wane. Years later, as a full-fledged surgeon, he returned with his wife, Nancee. This time the visit coincided with another move, one to a larger facility staffed with more personnel.

In 1958, Epp Memorial Hospital's nursing staff consisted of three single nurses: Canadian missionaries Mabel Ayer and Joyce Moe and María Luisa Granja, an Ecuadorian from the coast. Aside from operations that involved all three, each nurse had a particular field of duty: Mabel was responsible for patient care, Joyce did the sterilizing, and María Luisa assisted the doctors in the outpatient clinic.

Hospital records show that the first surgery in the new facility was performed on July 17, 1958. That morning new missionary Dr. John Wood gave Shiwango Andi a spinal anesthetic. While Liz Fuller checked the patient's vital signs, her husband assisted Dr. Wood in repairing Shiwango's hernia. Joyce passed the instruments, and Mabel circulated. The day ended with Sofia Salazar delivering Epp Memorial Hospital's first recorded baby, an 11-pound, 2-ounce boy.

The new hospital's unfinished second floor became a carpentry shop staffed by already overworked personnel. Nurses and any other staff available spent their evenings sandpapering and varnishing furniture as it was built. With only two beds and one bedside table, admissions had to be limited. While the nurses waited for the men to build the bedside stands, they heaved the same heavy table from bed to bed as needed. During the hospital's first year of operation, two carpenters made 14 wooden beds, 14 bedside tables and 40 wooden chairs.

Three months after the first patients were admitted, two local girls, Luz María and Angelita, asked to be added to the staff, and the hospital hired them as "nurse's aides in training." Liz Fuller, who instructed them, had no problems with Luz María, a graduate of the Berean Bible Institute. Angelita, on the other hand, could not abide by the hospital's rules and was asked to leave.

When it became plain that Luz María planned to stay, the Shell Mera staff asked the Palmer School of Nursing to grant her a certificate. (This School was opened by Rimmer Memorial Hospital in Quito in 1955 to prepare Ecuadorian women to become nurse's aides.) But Luz María had not had opportunity to attend formal classes, and she lacked the theory that Quito's year-long course provided. Mission representatives from both hospitals spent hours discussing her qualifications and hospital policy. In the end, Luz María was granted a certificate--the only time such a concession was made.

Unable to provide 24-hour nursing care, Epp Memorial Hospital staff hired Agustín and his cousin Alberto, two Indian boys from Dos Ríos, to take turns sleeping in the hospital at night. When Agustín was 13 years old, he accepted Jesus Christ as his Savior. His father strongly opposed his son's decision to become a Christian and threatened to kill him. In response, Agustín left home and went to Shell Mera to attend a Christian camp. Then, in May 1958, he accepted a position at the hospital where he remained until his retirement 35 years later. During the day, Agustín did maintenance work while continuing to sleep alternate nights in the hospital. Of those early years he says, "I didn't really mind the work; I was young and everything seemed exciting."

Each evening staff dragged woven straw mats out of a closet and placed one on the floor beside each patient's bed for a relative to lie on. Each relative was given a flashlight and instructions to waken the night boy if the patient needed a doctor or nurse. If the doctor on call lived in army housing, Agustín took his flashlight and ran across the airstrip to waken him.

Canadian missionary David Findlay was in charge of the hospital's administration and maintenance. The threat of fire constantly concerned Dave, who hung signs forbidding the use of matches, candles and small cooking stoves. While the signs were a good idea, they were not totally practical, for not all the patients and their relatives could read. One cold night, a woman innocently expressed a wish to build a fire in a corner of the room to keep her feet warm. This was dismaying enough, but when staff members discovered a charred area where an overturned candle had scorched a wall, Dave took further action. He and Agustín mounted an innovative fire hose that wound around a huge wheel. The homemade apparatus hung on the wall near the nurses' station; in response to a hearty yank, the hose unwound. A similar piece of equipment hung on the second floor. For the ongoing education of the staff, a fire drill was held once a year.

Since the hospital had no kitchen, patients' relatives carried into their rooms food bought at a local café and transported in a three-tier enamel pot.

The bottom tier held charcoal to keep the food warm; the second tier contained boiled rice and fried egg or *plátano* (cooking banana); and the top tier held soup. People ate the same food at each meal, without variation. After a few years, however, this system proved unsatisfactory, in part because not every patient was accompanied by a relative. Therefore, until a hospital kitchen could be built, staff decided to operate a kitchen on concession.

The hospital administration provided a suitable room in an outbuilding and hired Melba Urrutia, a local woman, to cook.

She was to be paid a set amount for each meal she served. This system worked fairly well until *Señora* Melba decided to increase her profits by raising pigs in the mission complex. She fed the pigs the leftover food from patients' trays and allowed the filthy and rowdy animals to root in the mud and grass where they gave a disheveled appearance to the hospital grounds. To relieve the hospital of this embarrassment, Dr. Art Johnston warned *Señora* Melba that if she did not remove her pigs, he would shoot one.

Though Art repeated his warning several times, the pigs remained, for *Señora* Melba was confident that the quiet-spoken doctor would never follow through on his threat. Finally, in exasperation, Art actually did shoot a pig. What a commotion! In her fury Melba, ranted and raved, bewailing her financial loss. In the end, Art paid for the pig, but Melba refused to give the meat to him or to the hospital. Instead, she pocketed the doctor's money and then took the pork into town and sold it. She was quite pleased with her earnings, but the incident did put an end to raising pigs on hospital premises!

As with all beginnings, a smoothly running hospital did not emerge overnight. Adjustments and improvements came slowly, but they did come, and the years would prove the worth of this new facility. Once the hospital was fairly launched, the concern uppermost in the minds of mission leaders became the question, "Will the Lord provide enough doctors and nurses to staff two hospitals and the Palmer School of Nursing in Quito?"

10
Perils on the Road

"Do not be afraid, for I am with you."

–Isaiah 43:5a

The plane arriving from Panama began its descent to Quito's Mariscal Sucre airport. In the crisp early morning air, snow-capped mountain peaks glistened in the sunlight. To the west the green slopes of Mount Pichincha stood out against a blue sky. From the plane's window the panoramic view was magnificent. My heart pounded with anticipation. This was Ecuador, the land of eternal spring. And this was Quito, its capital. Cradled in Mount Pichincha's lap, 9,600 feet above sea level, close to the line of the equator, Quito is one of the oldest and most beautiful cities of the Americas.

I was not alone on my first flight into Quito. With me were four other missionary nurses, three of them Canadians and two Americans. We were new recruits for HCJB's health care division. Its staff eagerly awaited our arrival from language school in Costa Rica where we had remained longer than the usual eight months, studying Spanish and waiting for permission to enter Ecuador. Although the Ecuadorian government was granting immigrant visas to other professional people, they had not granted them for us five nurses.

Despite our excitement, we were concerned for our future seemed uncertain. A consul had stamped our passports with a courtesy visa, valid for only 90 days, and on the same page he had clearly written the words, "This visa does not give you permission to practice your profession in any way." Our date of entry was August 5, 1959, a Wednesday.

Early on Sunday afternoon, August 10, in celebration of Ecuador's Independence Day, we mingled with the crowd of bystanders lining Quito's 10th of August Avenue to enjoy the pomp and splendor of a military parade. When it ended, Dr. Paul Roberts called us to the hospital.

Earlier in the day, Galo Plaza, a former president of Ecuador, had brought his critically ill mother to Rimmer Memorial Hospital and asked that she be cared for round-the-clock by American nurses. Dr. Roberts explained that although the hospital had the nurses, they did not have the government's permission to practice their profession. The eminent statesman replied, "You take care of my mother, and I'll take care of the nurses' licenses."

In spite of all the professional nursing care that Rimmer Memorial Hospital could provide, *Señora* Plaza died several days later. After her death, Dr. Roberts suggested that we visit the mission's medical work at the edge of the rain forest, since it might be some time before we were granted permission to work.

Early in the day we caught the bus for the long, tiring journey to Shell Mera. After rumbling along a cobblestone highway for some distance, our bus, in low gear, chugged slowly up the mountainside to the top of the *páramo* where we had an excellent view of Mount Cotopaxi's snowcapped peak. The main road snakes across the eastern range of the Andes' duplicate mountain chain. Crossing Ecuador from north to south, the eastern and western *cordilleras* are connected by numerous transverse ranges. Dotting the soaring summits are several snowcapped volcanoes, most of them inactive.

It was 4:00 p.m. when our bus reached the town of Baños. The driver announced a 20-minute stop, but before the weary travelers could buy the usual *fritadas y choclos* (fried pork and cobs of corn), he was honking the horn for an immediate departure. As the last of the passengers clambered aboard, stumbling across bags of produce lying in the aisle, the driver swung onto the street and roared through town.

Startled by this sudden departure, I turned to the passenger sitting across the aisle and asked, "Why is the driver in such a hurry?"

"It looks like rain," came the reply, "and he wants to drive through the river at Río Blanco before the water rises."

Out on the main thoroughfare, our bus continued full speed ahead, clouds of dust billowing behind. In the valley below, threatening rain clouds darkened the sky. To guide the cumbersome bus around the many dangerous turns, the driver constantly swung the steering wheel, first in one direction and then in the other. When he reached Agoyán Falls, he slowed and turned on the headlights before the bus entered a narrow tunnel only wide enough for a single vehicle. Moisture oozed from the tunnel's rocky walls, forming puddles on the road. At the far end, a speck of daylight showed. To warn oncoming traffic, the driver sounded the horn. Its blare resounded in the hollow passage until the bus made its exit. Emerging from the tunnel, it turned sharply to the left, clinging to the cliff as it wound around the mountainside. In that frightening instant, I had my first glimpse of the canyon's magnificence. Steep slopes and sheer precipices confine the raging Pastaza River. Though the deep drop-offs were terrifying, our bus sped on. We were within sight of the river, which we would have to cross at a ford when the bus jolted to a sudden stop.

To assess the problem, the driver asked the passengers to leave the coach. Commuters living nearby set off to complete their journey on foot. They could reach home faster by walking than by waiting for another bus. We passengers destined for Shell Mera plodded across the road in the sticky mud and scrambled up a grassy embankment in the pouring rain, leaving the driver to try to repair his motor. But the damage was major: a broken driveshaft. Returning to the broken-down bus, we waited for other transportation.

Soon a jeep drew up. Its driver offered to take as many passengers as he could. Three of us five nurses climbed aboard, leaving Beth Huddleston and me, along with two or three others, with the bus and its driver. Since no other traffic was moving in either direction, we were stranded.

After what seemed like hours, the driver of a car going to Ambato agreed to notify the depot that a replacement bus was needed. The afternoon dragged by as we endured another long wait. At the equator, darkness descends without warning, like a dropped curtain. Suddenly, the sun disappeared behind the Andes, leaving us hemmed in by steep mountain slopes, uninhabited and forbidding. As the minutes passed, I became more uneasy, but Beth remained calm.

I first met Beth in 1952 when I made a visit to the operating suite of Toronto Western Hospital, where we both worked. I wanted to invite her to a graduate Nurses' Christian Fellowship (NCF) meeting. I found her checking an anesthetic machine. A slight, blonde nurse immaculate in a white uniform, Beth, for all her quick movements, appeared aloof and reserved at this first meeting, yet she accepted my invitation. It was the beginning of a lifelong friendship.

As long as Beth can remember, she wanted to be a nurse. One day during her nurse's training, a colleague asked, "Beth, what are you going to do when you graduate?"

"I think I'll be a missionary nurse," Beth replied.

Beth's parents regularly tuned in to Sunday-morning religious broadcasts, and these programs had given Beth a desire to know the Bible better. To this end, she had written to Toronto Bible College for an application form. One of the requirements for admission to the Bible college was to describe a personal experience of salvation in Jesus Christ. Because Beth lacked assurance of salvation, she never completed the application.

Some time later, during a home Bible study, the simple gospel message became a reality to Beth, and the experience gave her an even keener desire to study the Bible. Within two years she began to turn her dream into reality by attending evening classes at Toronto Bible College. After another year Beth resigned from her position as an operating room supervisor to enroll in the college's day school.

One day in 1956 I read an article written by a reporter who had traveled to Ecuador after the killing of the five American missionaries. His description of a 35-bed hospital operated by a mission in the city of Quito impressed me. Thinking that Beth would be interested in this hospital, I gave her the article to read. She scanned it, but to my disappointment, she made no comment. Why, I wondered, was she not interested? Had I been wrong?

One evening, after we had attended the college's missionary conference, Beth said, "I feel that the Lord is directing me into foreign service, but I'm not sure where."

"Doesn't one country come to your mind more than any other?" I asked.

To my surprise, Beth answered, "Yes, Ecuador."

Unknown to me, Dr. Paul Roberts had spoken at the college chapel one morning. After the service, Beth had sought him out to learn more about HCJB's medical work. Now, months later, after much prayer, she applied to HCJB World Radio and was accepted. In His own way, God had directed her to Ecuador. She never did remember the newspaper article I had given her to read.

In the evenings Beth shared what she was learning at Bible college. Her joy in the study of the Scriptures was contagious, and soon I, too, began to hunger for more Bible knowledge. Thus, after 13 years on the staff of Toronto Western

Hospital, I resigned from my position as head nurse to begin a year's study at the same college.

Mission work in Ecuador was uppermost in my mind, but how could I be certain that it was the Lord's will for me? I did not want to go to Ecuador just because my best friend was going. One Sunday, like Gideon of the Old Testament, who put out a fleece to determine the Lord's direction and found his fleece wet when all the ground around was dry (Judges 6: 37-39), I asked God to do something similar: to bring Ecuador to my attention in unexpected ways every day for one week.

On Monday evening I attended a Nurses' Christian Fellowship meeting. Since the speaker, Muriel Stewart, president of the group, had just returned from visiting her sister on a mission field in Africa, I was expecting to hear about the work there. To my surprise, Muriel talked about Ecuador. Having read the recently published book, *Through Gates of Splendor*, she based her message on the dedication and commitment of the five men who had tried to take the gospel to the Aucas.

Surprised that the Lord had answered my request the first day, I was even more amazed when a hospital technician stopped me on Tuesday morning and said, "I hear that you live with Beth Huddleston." I had heard Beth speak of Myra and Dixie Dean, who were attending Bible college with Beth and were accepted candidates with HCJB. I had not known, however, that they were this technician's sister and brother-in-law. For the next several minutes Myra's sister told me all she knew about Ecuador. In similar ways the Lord brought Ecuador to my mind each day, assuring me that He wanted me to serve Him there.

Later, when I, like Beth, was accepted by HCJB, I had heard Clarence Jones say to a number of us missionaries in the orientation stage of our training, "It is always good to remind yourself of the way in which the Lord has led you, for on the mission field you will experience difficulties that will cause you to doubt." I was reminded of his words as we sat in the darkness of the gorge, waiting for another bus. I did not doubt the Lord's guidance, but I had not overcome the problem of fear.

Our small group of stranded passengers waited two hours before a replacement bus arrived from the depot. As this bus lumbered through the stony river bed at Río Blanco, rushing water from the rising river swirled around its wheels. For the next two hours we drove along the tortuous route through the gorge of the Pastaza River. To the right were treacherous precipices, some of them dropping hundreds of feet. Darkness added to the dangers of the seldom-traveled road, unnerving even the best of travelers, and I was no exception. When we finally arrived in Shell Mera, I announced to Beth, "We'll have to find another way out of this place. I am *not* traveling that road again!"

Two days later Mission Aviation Fellowship had a flight leaving for Quito with two empty seats. While our three colleagues traveled overland, Beth and I went by plane. It was my first flight in a small aircraft, and I was nervous. Trying to reassure me, Beth said, "Close your eyes and don't look at the mountains--and remember you'll not die one minute before your time."

When we returned to Quito, we received permission to begin nursing at Rimmer Memorial Hospital. Within a few months, and in answer to our prayers and those of the HCJB staff, our passports were stamped with an immigrant visa valid for an indefinite time. But before we could celebrate, a dreadful accident occurred on the Shell-Baños road.

On Sunday, November 15 a stranger rang the doorbell at Epp Memorial Hospital, bringing the tragic news that a local bus on its way to Quito had careened over the cliff at Cashurcu, near El Topo. Jungle foliage closed over the wreckage, hiding it from sight. The tragic accident might not have been discovered for days, except that a young lad riding in the rear of the bus was thrown out as it went over the cliff. He landed in the tall grass along the roadside. He scrambled onto the roadway, flagged down the next car that passed and told the driver what had happened.

As soon as the word reached Shell Mera, hospital staff climbed into the mission's jeep and headed up the road. From the tracks that the wheels of the overloaded bus had made, it appeared that it had swooped down a steep hill, struck the rocky wall on its right, and then veered across the road and over the cliff. Its brakes had failed at a point where the cliff dropped a sheer 300 feet. By now rescuers were bringing injured survivors along the riverbank to a path that led up to the main road. Staff returned to the hospital to prepare for the arrival of the casualties.

News of the tragedy spread quickly. Soon a crowd converged on the hospital. Desperate to locate a missing family member, relatives and friends--some hysterical--forced their way inside, checking each survivor. It was impossible to keep the grief-stricken relatives out of the emergency room. Within minutes the little hospital had become a disaster relief center. Since the hospital had only 14 beds, some already occupied, nurses placed accident victims on mattresses on the floor and knelt to attend them. Inpatients who were well enough were discharged to nearby homes to make room for new admissions. To maintain order, and to console the bereaved, teachers and students from the Gospel Missionary Union's Bible Institute assisted the staff.

Of the persons on the bus, 33 died and 13 were injured. Among the survivors was the bus driver, Segundo Bonilla, whose leg was badly fractured. The doctors took him to the operating room and inserted a metal pin in the broken bone. When the surgery was finished, they slid their hands beneath the semiconscious Segundo to move him gently onto a stretcher. Then, to their chagrin, they discovered another fracture in the same leg; it had been overlooked during the initial examination.

Early the next morning the doctors were on the radio contacting Quito. They desperately needed an orthopedic specialist, a longer metal pin for Segundo's second fracture and more nurses to handle the critical patients. Dave Findlay contacted Gonzalo Ruales and asked him to fly to Quito to bring back the extra personnel. Dr. Guerra and Dr. Uquillas, Ecuadorian orthopedic specialists, canceled their office hours and, with two missionary nurses, Mabel Ayer and Billy Hoboth, flew to Shell Mera. The visiting doctors came equipped with the necessary orthopedic instruments to perform the open reduction on Segundo's fractured leg.

To avoid charges, Ecuadorian drivers involved in an accident try to flee the scene, for a driver who can escape and hide for six months usually goes free. Lying in hospital under police guard, the reckless Segundo regretted that his broken leg prevented flight. Encased in a cast and feeling frustrated, he became a demanding patient, thinking only of himself. He expressed no concern for other patients who were also suffering, nor did he show any remorse for the lives that had been lost.

Most of the victims were local people, but some patients lived as far away as Tena. One survivor was *Señor* Sevilla, on whose hacienda Rachel Saint had learned the Waorani (Auca) language from Dayuma. Having lived in Auca territory for many years, *Señor* Sevilla had survived wounds inflicted by Aucas. Rescued from the bottom of a steep cliff, he made another good recovery under the care of Epp's staff.

Following the accident, a dark cloud of grief hung over Shell Mera and Puyo where several families had suffered the loss of loved ones. Survivors' accounts of the accident were emotional and heartrending. "Why am I still alive?" one pregnant woman asked a nurse. Sitting in the middle of a seat for three, the woman had watched both her seatmates die.

Hoping to help their patients, staff returned to the hospital during evening hours to read Scriptures and share their faith with the injured, yet they saw no immediate fruit of their labors. Despite a near-death experience, none of the survivors made a commitment to follow Christ. Doctors and nurses could only hope and pray that their patients would respond to the gospel at some future time.

Three months had passed since the five nurses who made up our group had traveled to Shell Mera with the driver whose bus, in continual disrepair, now lay crumpled at the bottom of the cliff. It was a sobering thought that, but for the Lord's protection, our trip could have been the fatal one. At all times the Shell-Baños road was treacherous. No guardrails protected travelers at sharp curves and steep precipices. Some bus passengers felt safest riding with the chickens and stalks of plantain on the roof so that they could be ready to jump off in an instant. Yet these were the people who plunged to their death the moment the bus went over the cliff.

The long, tiring and dangerous Shell-Baños trip kept most missionaries at home. Fearful of mountain roads, Mabel Ayer went to Quito only once during the 15 months she spent at Epp Memorial Hospital, and after she transferred to Quito, she rarely found the courage to return to Shell Mera. When she did make the trip, she kept her eyes closed to avoid seeing the dangerous drop-offs. I, too, feared the road. During my next 12 years in Quito, I would make infrequent visits to the edge of the rain forest.

Some staff members, however, were much more courageous. During one all-night drive from Shell Mera to Quito, the headlights of Dr. Johnston's car failed. Undaunted, he and his two passengers snaked through the Pastaza River gorge and across the Andes Mountains by the light of a full moon!

11
Surgery by the Book

"For the Lord gives wisdom, and from his mouth come knowledge and understanding."

–Proverbs 2:6

The road that winds through the canyon between Baños and Shell Mera was deserted and lonely at night. For the local people, day began before sunrise and ended at sunset. Shrouded in darkness, their drab houses stood silent by the side of the road. Since everyone traveled by bus, there was very little traffic on this hazardous dirt road.

It was late afternoon when Dr. Ev Fuller and Hospital Administrator Dave Findlay finished shopping in Ambato and began the long drive down the mountainside to Shell Mera. Just beyond Río Verde (Green River) they spotted some people carrying a man on a makeshift stretcher. Dr. Fuller stopped the car and asked, "Can we help you?"

These were welcome words to the little group, for Pedro, the man on the improvised stretcher, was badly injured. Moving boxes and parcels to one side, Ev and Dave quickly cleared the rear of the van to make room for Pedro and his wife, Rosa. During the next two hours as they drove on to Shell Mera, Ev and Dave learned the cause of Pedro's accident.

Rosa and Pedro lived in a small frame house on a hillside near Río Verde. When the accident happened, Pedro had been cutting wood. In the jungle, undergrowth--much of it vines--

often leaves a woodsman cutting trees no room to make a horizontal swing. Instead, he must bring his ax or machete up over his head, whirling it in a continuous motion as he works in the confined space. The tree Pedro was cutting fell suddenly, striking him on the head, throwing him against the sharp stump, and damaging his face beyond recognition.

When the small group reached the hospital, Dr. Fuller quickly alerted the staff. That night everyone returned to the operating room where Dr. Espinosa, the dentist from the military base, joined them at Dr. Fuller's invitation. Then Dr. Fuller and Dr. Johnston began to repair Pedro's damaged face.

In the bright light of the operating room, the doctors were better equipped to see the extent of Pedro's injuries. His face--if one could call it a face any longer--was horribly mutilated. The accident had peeled away his nose and one eye, exposing the optic nerve as clearly as a diagram in an anatomy textbook. At the base of Pedro's skull was another gaping wound. The three doctors worked all night, cleaning the dirt from the wound and carefully piecing the torn tissues back together.

Twice during the night Dr. Fuller interrupted the operation with the order, "Time to cover up!" To ventilate the operating room, the upper half of its windows were screened, but tiny insects, attracted by the bright light, squeezed through the screens and landed on the surgical field. In response to the doctor's instruction, Liz Fuller, the instrument nurse, covered the open wound and her table with sterile drapes. Then the circulating nurse picked up the container of bug killer and moved about the room, pumping her spray gun to spread the toxic vapor over everything, including the window screens. This procedure was routine for all-night surgeries.

When the repair surgery was finished, Dr. Fuller did a tracheotomy to help Pedro breathe. Since Pedro's jaws were wired together, the doctor also inserted a tube into his stomach for direct feedings. Then he bolted through the swinging doors, covering the distance to the nurses' station in what seemed to be three long strides, shedding his surgical cap and mask en route. With each step he took, the hospital shook. At the nurses' station, he dashed off a few postoperative instructions in a barely legible scrawl, filling a page of Pedro's chart.

Although Pedro made a good recovery, his accident and the surgery that followed left him with an unusual sensation that seemed to him like a tossing sea splashing against a rocky shore. Pedro described the feeling by saying, "There's something sloshing around inside my head." Since the doctors could find no cause for this strange sensation, they decided that Pedro was suffering from an iodine deficiency, a condition that was common in the Baños area.

Like Pedro, many of the patients whom the doctors attended at Epp Memorial Hospital were difficult cases. Yet it was too costly to fly them to Quito for special treatment, and it would be 20 years before Ecuador's government would provide an "air ambulance" service to help cover transportation costs for patients who could not afford travel for medical purposes. Thus the Shell Mera doctors found that their most valuable tool was the surgical textbook.

Time and time again it found its way into the operating room. Doctors peered over their masks at various diagrams in the book placed on a stool beside them and open at the right page. When they wanted to read a passage, the circulating nurse held the book at a readable distance.

Operations that go smoothly and without complications are seldom remembered. The unusual cases are the ones that staff recall. One such case was Ulrich, a German-speaking patient whose arm had been severed when he stumbled into a circular saw. Expecting to find blood pulsating from the stump, the staff was surprised, on uncovering the wound, to find blood vessels tied off with grass. But in spite of the ingenuity of his co-workers who had arrested the bleeding, Ulrich's vital signs and blood test showed that unless the fluids he had lost were replaced at once, he would soon be in shock.

To obtain a unit of blood was not easy. Shell Mera's supply came from a "walking blood bank," usually the foreign staff. On this occasion, GMU missionary John Stuck, who was passing through Shell Mera on his way to the jungle, volunteered to be the donor. The morning after the transfusion, John visited Ulrich and explained that he was particularly interested in his progress because he had donated his blood. Overwhelmed by John's generosity, Ulrich exclaimed, "Thank you! Thank you! Now I can speak English!"

Taking the opportunity to make the gospel message known, John replied, "Someone else gave not only His blood, but also His life for you."

Sometimes the gospel message is so new to hearers that it is hard for them to grasp it. The sacrifice that Christ made at Calvary was a completely new concept to Ulrich, who had never heard the story before. During his time in the hospital, where the medical work was closely tied to evangelism, he would hear more of the story of Jesus, for engineers had installed radios pretuned to Radio Station HCJB to provide spiritual teaching in the patients' rooms. Hospital chaplains find more opportunities to explain the Scriptures to patients who listen to programs broadcast by a Christian station.

Of necessity, Epp Hospital staff learned to be creative and resourceful. Dr. Art Johnston was especially ingenious. Sometimes he would disappear before an operation. While the nurses waited and wondered, he would be in the carpentry shop making an apparatus needed for the patient's postoperative care. When he finally arrived for surgery, he would have the most unusual invention--not standard hospital equipment--but something that certainly served the purpose. At other times, to the horror of the surgical nurse, he would bring her a greasy old tool and ask her to clean and sterilize it for the surgery he was about to perform.

When Dr. John Wood left Shell Mera, Art Johnston inherited one of his colleague's patients. Piedad Salazar, a local woman, had recently had an emergency operation for a perforated appendix. As sometimes happens, the incision would not heal. In a medical article Art had read, "If a wound will not heal, apply placental tissue to the area. This will stimulate new tissue growth." Art decided to try the procedure, and to his amazement, the wound healed, though it left Piedad with a weird-looking scar that resembled a sunburst. She never complained, however, about her ugly scar, since she had no plans to wear a bikini.

Dr. Johnston's ingenuity came to the fore the day he admitted Luis Vargas, an Indian chief. Luis, a paraplegic, suffered from Pott's disease, an inflammation of the vertebrae caused by tuberculosis. To promote the young man's recovery, Dr. Johnston wanted him placed in a Stryker frame so that the nurses could turn him easily. The hospital had no such equipment, but after sketching a diagram supplemented with specific instructions, Art asked Angel, the hospital carpenter, to assemble the complex device.

Angel met the challenge by using two wooden stretchers to build a double frame with troughs for Luis' legs that extended from his feet to his armpits. When it was finished, the upper and lower parts of the frame were clamped together with Luis sandwiched between. The contraption provided a rest for Luis' forehead and a hole for his face when he was lying face down, allowing him to eat his meals by reaching around the edges of the stretcher to his food tray, which was placed on a stool below. It also provided a hole large enough to insert a bedpan that Luis could use when he was lying on his back. Angel was proud of his apparatus, which he described as a wooden box resembling a coffin. His description was apt in more ways than one, for each time Luis needed to be turned, the nurses had to recruit the two heftiest people in sight to rotate the frame.

During the months he spent in the hospital, Luis received medication for tuberculosis. When his condition improved, the nurses removed him from the homemade frame and helped him to his feet. Shuffling one foot ahead of the other, Luis moved along the hallway with a spastic gait. Though after such a lengthy illness rehabilitation would take some time, the joy on Luis' face was equaled only by that on the faces of the staff as they saw their patient on his feet again.

The doctors and nurses were often amazed at how well people in the area managed to improvise. One day while Miguel Bustamante, a homesteader, was clearing his land, he dropped his machete on his foot, severing a number of tendons. Alone and miles from medical help, the injured Miguel used the only medication available. Taking cold wood ashes from his bonfire of the day before, he packed them around the wound and bound his foot with his dirty shirt. Loss of blood and pain from the injury forced him to rest, but when he regained enough strength, he whittled a crude crutch and hobbled through the woods to the main road. There a passing truck driver picked him up and brought him to the hospital. When Dr. Johnston removed the dirty rag covering Miguel's injured foot, he expected to see a terrible wound; instead, it was remarkably clean. Lye in the ashes had worked as an antiseptic, preventing infection and promoting healing.

Sometimes it seemed that the difficult cases arrived at the hospital late in the day or early in the evening so that the day staff had to return to work at night. *Señor* Vega, mayor of Puyo, and two of his fellow townspeople would not have needed medical attention at night if a small boy carrying a lighted candle had not stopped to watch him as he poured alcohol into a container. The inquisitive lad's candle ignited the combustible liquid, burning not only the boy himself, but also *Señor* Vega and a woman standing nearby. When the victims of this accident arrived at the hospital, they were in critical condition. Both the woman's burns and those of the young lad extended from head to foot; both patients died before dawn.

Art and Verna Johnston took *Señor* Vega to the operating room where they worked all night, cleansing the burned areas and applying sterile dressings. Sixty percent of *Señor* Vega's body was burned, a good indication that he might not live. Yet the man had unusual stamina, and the presence of a caring relative encouraged him. Noticing that his sister came to the hospital and spent long hours at her brother's bedside, nurses took the opportunity to instruct her in his nursing care. She learned well and cared for him almost constantly as he slowly recovered from the burns and several skin grafts that followed. *Señor* Vega benefited from his sister's support, since the shock of a severe burn, with its resulting disfigurement, is so traumatic that patients need intense emotional support from family members. When he was well enough to return to his duties as mayor, he never forgot his friends at the hospital and used his influential position to help them whenever he could.

During the next 27 years, much of the drama at Epp Memorial Hospital would take place behind the closed doors of the operating room where hospital surgeons, visiting specialists and general practitioners would perform more than 6,000 operations. Yet not all of the hours spent in the operating room were stressful ones. The Shell Mera staff did enjoy some lighter moments. If Art Johnston was scrubbed, he was sure, it seemed, to discover a flea in his sock. Not wanting to contaminate his sterile gown and gloves, he would prance around the table, exclaiming, "Will someone please get that flea!"

Holding a glass of water to drown the flea, the circulating nurse would crawl under the table to search the doctor's ankle for the little pest. Fleas had an affinity for Art, though they never seemed to be attracted to anyone else. He never relaxed until the nurse showed him the glass of water with the submerged flea. Thus, despite their isolation, staff in Shell Mera maintained a sense of humor, and outside of working hours they enjoyed some fun times, too.

12
A Shell Mera Romance

"The Lord God said, "It is not good for the man to be alone."

– Genesis 2:18

Dr. Ev Fuller, a leader in the field of medicine and highly respected in the community, quickly distinguished himself throughout the *oriente*. Keenly interested in sharing medical knowledge, he soon established rapport with the army doctors at the Shell Mera base. He invited them to assist him with interesting surgical cases, and they reciprocated. Not only did Ev and Liz build relationships with the military personnel, but they also identified closely with the civilian population. They knew everyone in the village and mingled with them, regardless of the local people's view of social status.

For relaxation, Dr. Fuller and Dr. Espinosa, the military dentist, who were both tennis enthusiasts, often spent Saturday afternoons playing tennis at the army base. While the two men were on the courts, their wives and the single nurses played table tennis, chatted and drank coffee. In addition to these social outings at the army base, Epp Hospital staff often picnicked at the Alpayacu River, a popular place to swim and relax in the sun. Loading everyone into the truck, including Storm, the family dog, the group would drive the two miles along the bumpy road to the river. Invariably, Storm vomited along the way.

Since flying in the *oriente* is scheduled for daylight hours, another diversion was to stroll about the airstrip in the evenings. In fact, young Dave Findlay chose to court Joyce Moe on Shell Mera's gravel runway, realizing that there was nowhere else where he and Joyce could avoid the scrutiny of the senior missionaries and the Ecuadorian staff. Shell Mera did not even have a decent restaurant where the couple could sit and drink coffee. Thus, during the evening, the unlit airstrip, noisy in the daytime but peaceful at night, became a lovers' lane.

While in Canada in the mid-1950s, Dr. Paul Roberts had invited Dr. Findlay, a dentist, to visit Ecuador with his son, David; young Dave had already completed three years at Moody Bible Institute, one year at Nyack Missionary College, and a year at the Missionary Health Institute in Toronto. Before his visit, Dave submitted his application to HCJB and was accepted for work on the field. Like most aspiring missionaries, he had some apprehensions. While he was visiting in Shell Mera, Dorothy Brown called him to the clinic to see a woman patient whose jaw was hanging open.

"Take a look at this," Dorothy said. "I'm wondering if it's tetanus."

Dave examined the patient's misplaced jaw and, remembering a similar case he had seen in Toronto, ventured his opinion, "I think it's a dislocated jaw."

"What should we do" Dorothy asked. "There's no doctor here."

Dave volunteered to try to adjust the jaw. When the woman had been seated on a chair and Dorothy had administered just enough trilene anesthetic to keep her comfortable, Dave put the jaw back in place. Dorothy was so impressed that she exclaimed, "This is where you ought to be!"

It was certainly a positive experience for patient, nurse and "doctor." At the end of the Findlays' visit, Dave and his father returned to Toronto, where Dave made preparations for full-time missionary service. After several months of language study in Costa Rica, he returned to Shell Mera. When he arrived, Dr. Fuller asked him if he would be interested in learning to give anesthetics. Dave was eager, and so, in addition to his administrative duties and maintenance work, whenever the doctors needed him in the operating room, Dave left his routine work and put on surgical greens.

Joyce Moe's interest in the medical work at Shell Mera began with a letter from Dorothy Brown, her friend from Bible college days. On one page, Dorothy printed, in large block letters, the words, "WE NEED NURSES." She went on to say, "This is actually an HCJB hospital, but Gospel Missionary Union has promised to supply three nurses to help out here. Come with either mission." Then, before sealing her letter, she slipped Dave Findlay's prayer card and picture into the envelope with the note, "If you don't come to the jungle to do medical work, you could come for this reason."

Joyce laughed when she read Dorothy's message and put the letter back in its envelope. She was uncertain where the Lord was leading her, but from their recent correspondence, she knew that Dorothy was anxious for her to join the Shell Mera team. Some time later Joyce wrote to both missions for an application form, deciding to fill out the one that arrived first. It was GMU's. She filled out the application and sent it to the mission. In response, GMU asked her to come to its headquarters for orientation. There Joyce met HCJB's Bob Savage, who said to her, "If GMU doesn't accept you, HCJB will."

GMU did accept Joyce and sent her to Epp Hospital where she took charge of the operating room. She always arrived early to prepare for surgery, and Dave came ahead of time to check his anesthetic equipment. This gave the young couple a few minutes to be alone. One day Dave took the opportunity to invite Joyce for an evening stroll on the airstrip. To avoid drawing attention to their romance, the couple kept changing their meeting place; sometimes they met at the northwest end of the runway and sometimes at the southeast end!

Those evening strolls were unforgettable. On clear nights, moonbeams filtered through the branches of the jungle forest to cast a silvery glow on the snow-covered peaks of El Altar and Sangay. The enchantment of the moonrise cast its spell on the young pair, and love blossomed.

Apart from working hours, Joyce and Dave had little opportunity to be together, but occasionally they enjoyed a day's outing at Baños. The drive there took two hours or more if one stopped to enjoy scenic spots along the way. On these outings, Dave's and Joyce's favorite eating place was the Gertrudes Hotel where the owners watched the romance develop.

The young couple planned a June wedding, scheduling a morning ceremony so that they could fly to Quito in the afternoon to begin their honeymoon. The week before the wedding, Dave had booked reservations at a hotel in Quito and had sent their suitcases to the capital city by truck. Since Gonzalo Ruales had loaned Bob Wittig his Cessna for the occasion, Dave arranged for Bob and Keitha Wittig, now living in Quito, to bring the wedding cake and flowers when they flew into Shell Mera for the ceremony. They would arrive in Shell Mera in the morning, and the newlyweds would fly back with them to Quito in the afternoon.

June is usually the rainiest month in the *oriente*, and June 1959 was no exception. On the day of the wedding Shell Mera's weather was dull and cloudy. It was a beautiful day, however, in the capital city of Quito, and Bob and Keitha started out early for Shell Mera. When they reached Baños, thick clouds filled the gorge, and Bob flew back to Ambato's landing field and waited. Forty-five minutes later he tried again, but the weather at Baños was still cloudy. Then, just before noon, the pass opened enough to allow him to make the flight. The hour for the wedding ceremony was now changed to 1 p.m.

Dave Findlay and Joyce Moe were married in the chapel

of the Berean Bible Institute. The invited guests sat inside, while the village folk, brimming with curiosity, peered in the open windows. No one wanted to miss Shell Mera's social event of the year.

In the bridal party were two missionary children, Jenny and Jeff Wood. The minister had begun to pray when Jenny's mother noticed that her little daughter was shifting restlessly from one foot to the other. Realizing the problem, she slipped out of her seat, took Jenny from the wedding party and led her outside to the bushes nearby. Fortunately, the opening prayer was long. When mother and daughter returned to the chapel, the pastor was still praying, and little Jenny tiptoed quietly into her place, her absence unnoticed.

When the ceremony ended, and Dave and Joyce were hurrying from the chapel, Luz María's mother, *Señora* Cárdenas, thrust a live chicken into Joyce's hands. Joyce was completely baffled. What was she expected to do? She didn't want to offend the *señora*, for in Ecuador a chicken is an expensive gift for a poor person. Neither did she want to soil her dress, which she had borrowed from a friend in Quito. Realizing the predicament her mother's gift had caused Joyce, Luz María whispered, "Can I relieve you of that chicken?" Joyce promptly handed her the live bird.

By the time the reception was over, Shell Mera's weather had closed in, and it was impossible for a plane to take off for Quito. Sensing the dilemma of the bride and groom, two friends offered to drive them in their jeep as far as Baños. The bridal pair gladly accepted.

At Baños the newlyweds headed for their favorite spot, the Gertrudes, but to their dismay, they saw the car of one of their wedding guests parked outside the inn. He had arrived just ahead of them. Deciding that they preferred to stay elsewhere, Dave and Joyce went on two blocks down the road, where they found the only other accommodation available: a room furnished with two single beds with straw mattresses.

The next morning Dave talked with Bob Wittig by ham radio. Learning that the weather was clear in the *oriente*, he arranged to meet Bob at the airport in Ambato in one hour. After a search, Dave located a bus, the only one available. It was an open jalopy with a hole in the floor and a door with a missing hinge. Once the passengers had boarded, the driver wired the door shut, and the bus started off. The honeymooners bounced along the dirt road, breathing in clouds of dust that rolled up through the hole in the floor and covered them from head to foot. Though the rattling bus took the Findlays safely to Ambato, it was the most dilapidated one in which they would ever travel during all their years in Ecuador.

Back in Shell Mera after their honeymoon, Dave and Joyce settled into one of the two apartments on the second floor of the hospital. Whenever they were needed for night surgery, Agustín would use a wooden broom handle to knock on the ceiling under their bedroom. To improve this primitive system of communication, Dave and Agustín installed wooden telephones with crank bells. The system, which had 10 numbers with long and short rings, connected the missionary homes and Mission Aviation Fellowship with the hospital. The missionaries found it a great improvement, but Agustín, raised in the jungle where telephones do not exist, preferred to go on using the broom handle. It was sometimes more reliable than the crank telephone.

13
Hospitality

"Do not forget to entertain strangers, for by so doing some people have entertained angels without knowing it."

– Hebrews 13:2

The hour was late when Dr. Wally Swanson, with his family, turned the green jeep into the driveway of Epp Memorial Hospital. The diesel generator had been turned off for the night, and the place was in darkness; only the dim light of a flickering candle came from the Johnstons' apartment on the second floor. Art and Verna had not yet gone to bed; they were waiting for the arrival of the newest members of Shell Mera's missionary family.

Wally and Charlotte Swanson, who, with their three children, arrived in Ecuador from language school in August 1961, had settled into temporary living quarters in Quito. Six weeks later, Char gave birth to their fourth child. While she cared for their young family, her husband began to hold mobile medical clinics in Ecuador's northern region.

A short time later Dr. Ev Fuller traveled to Quito on business. He had been feeling a need for further training and had applied for a surgical residency in Panama. Arrangements for this training were now confirmed, and he hoped to leave Ecuador in December.

Wishing to take advantage of the opportunity to become acquainted with his colleague while he was in Quito, Wally invited Ev for dinner one evening. Seated on wooden crates in the Swansons' sparsely furnished apartment, Ev asked, "Wally, how would you like to replace me in Shell Mera?" Briefly he explained his plans to the startled Wally. "Tomorrow I meet with HCJB's board of trustees," he said, "and I'd like to be able to tell the board that you will replace me during my absence."

Although Ev's unexpected proposal surprised the Swansons, the Lord had prepared them for the change. Even before their arrival in Ecuador, the mission had vacillated about the Swansons' assignment, though the possibility of their working at the jungle hospital had not been mentioned. The Swansons agreed, however, to make the change. The board of trustees gave its approval, and two months later the young family moved to Shell Mera.

To allow the Fullers to attend the Christmas program at the Alliance Academy in Quito where their children were enrolled, Ev and Liz arranged to leave Shell Mera before daybreak. When they reached Quito, they left the hospital's green jeep with the Swansons to use for their trip to the *oriente*.

Wally Swanson's first journey to Shell Mera set a pattern for most of his future trips. It was late that afternoon when he finished packing the jeep, collected the mail for the staff at Shell Mera and ran last-minute errands. Leaving late in the day had some advantages: the long tiresome journey after sunset lulled the four children to sleep and gave their parents a peaceful trip down through the mountains. Seven hours after leaving Quito, the Swansons arrived in Shell Mera.

At the sound of the jeep in the driveway, Art and Verna Johnston hurried out to welcome the hospital's new family. Weary and hungry from their long drive, the newcomers were soon devouring Verna's fresh chocolate cake and drinking lemonade by candlelight. Then, since the hour was late, they crept silently into the frame house next door to the hospital. In the darkness the big house seemed spooky to the children, who imagined weird and frightening objects. When his parents whispered to four-year-old Jeffrey to be quiet because other house guests were in bed and asleep, he replied in a louder whisper, "Ugh! I'm not going to sleep here. There are bears!"

In fact, the big brown house contained no bears, only large, hard-shelled cockroaches that multiplied overnight, tiny ants that appeared from nowhere, and clever spiders that trapped their prey in intricate webs. Soon the Swanson children would discover that rhinoceros beetles and fuzzy tarantulas inhabited the trees outside their new home.

Wally had slept only a few hours that first night when he woke to the sound of someone beating on the bedroom wall with a broom handle and calling, "Doctor! Emergency!" The primitive telephone had failed.

"I'm coming," Wally called as he rolled out of bed. The room was pitch black as his feet touched the floor and he fumbled for the flashlight. When he stepped outside into the quiet night, the hospital was in darkness. Moments later the chugging of the diesel generator sounded, followed by a steady hum, and then the lights came on. Art and Verna were already in the emergency room. During Wally's next four years at Epp Memorial Hospital, the events of that first night would be repeated many times.

Verna spent most of the night in the operating room, assisting the doctors. Next morning she prepared a breakfast of scrambled eggs, sweet rolls--her specialty--and fresh pineapple for her own family and the Swansons. In order to give new staff time to adjust to their environment, senior members prepared meals for newcomers during their first three to five days in Shell Mera. Because Liz Fuller was gone, however, the doctors needed Verna to assist them in surgery. At supper time Verna prepared food and sent it to the Swansons, asking if they could then manage on their own. The Swansons' transition into life at Shell Mera was an abrupt one, for the hospital was full and its limited staff overworked. While Wally adapted to the practice of medicine at the jungle's edge, Char struggled with adjustments at home.

When the Swansons moved to Shell Mera, Char took over Liz Fuller's fully equipped home. Built before the completion of the hospital, the house was huge: it contained 14 rooms, seven of them bedrooms. Like the hospital, it sat on cement posts and had a roof of corrugated aluminum with overhanging eaves to protect its walls from torrential downpours. A covered veranda protected two sides of the house. A full-sized attic on the third floor served as storage space and provided a clothesline for use on wet days. In the still air under the hot aluminum roof, the linen dried stiff as boards.

In addition to sheltering the Fullers and their seven children, the house had provided accommodation for guests, working visitors, foreign medical students, boarders and the maid with her two children. Sometimes as many as 21 people had eaten their noon meal in the Fullers' home.

The furniture was where Liz had placed it; the linen closet contained a supply of sheets and towels; the kitchen cupboards were stocked with the usual staples; flour and sugar were in the bins, and the big freezer contained some blanched peas. The provisions were a help to Char in getting meals ready for her family in the new surroundings--but the next day three guests arrived. The guests were Dr. Murray Weaver, a distinguished surgeon from California, his wife, Lona, and Mr. John Harstone, director of HCJB's Canadian office. Dr. Weaver, a kind and gentle man with years of surgical experience, had applied to HCJB. Before beginning work in Ecuador, he wished to see the types of surgical cases encountered at the jungle hospital. The better to represent the mission in Canada, Mr. Harstone, a retired schoolteacher, was gathering detailed information from each department.

Verna, busy in the operating room, sent a message to Char, asking her to prepare for the guests. Char was overwhelmed. Her four children, ranging in age from two months to six years, needed considerable attention. Unable to plan a meal and shop on such short notice and in unfamiliar surroundings, she did the next best thing. Certain that she and Wally would not be returning to Quito before the end of the year, she had brought a smoked ham and other goodies to Shell Mera to use at Christmas. Opening these precious supplies, Char prepared a sumptuous meal. That evening, December 12, 1961, her family and guests sat down to an early Christmas dinner in the foothills of the Andes.

Since Dr. Weaver was tired from a long strenuous day in the operating room, he and Lona went to bed as soon as the meal was over while Wally returned to the hospital. The fastidious Mr. Harstone, sensing the frustrations of the new missionary mother, helped her wash the tall pile of dirty dishes and held the baby while Char put the other children to bed.

During his first term on the field, Wally Swanson, besides carrying out his duties at Epp Hospital, worked on his thesis to validate his medical license in Ecuador. Coordinating flights with the MAF pilot, he flew to jungle stations where he examined people of the Secoya, Cofán, and Waorani tribes and recorded his findings. Since the jungle stations had no electricity, the resourceful Wally took along simple battery-operated laboratory equipment. Using samples of his own blood as a control, he compared his blood with that of the Indians. Although the method was not perfect, it did show him that anemia was high among the bare-footed tribal people, among whom hookworm is common. The parasite larva enters the sole of the foot and becomes an adult worm attached to the wall of the intestine, where it sucks the host's blood, causing anemia. Wally was able to demonstrate increases in parasitic diseases and anemia since Dr. Fuller had made similar studies some years before.

All three tribal groups, Wally observed, had the same eating habits. They began the day by eating a banana and drinking *chicha*, and continued to eat throughout the day, having no fixed hour for meals. They ate a big meal only when the men returned from hunting with fresh meat.

The only birth-control method practiced among the three groups was abstention. Cofán parents abstained from sexual intercourse until their baby reached the age of 18 months or two years. In the Waorani tribe, abstention lasted even longer, usually until a child reached the age of four. Among the groups that practiced polygamy, the husband simply lived during that period with one of his other wives.

Wally obtained valuable information for his thesis while working with the tribal people. His data included the Indians' concept of hygiene, their attitudes toward the elderly and infirm and their use of jungle plants to treat common ailments. Travel to these outposts gave the young physician a deep appreciation of the medical needs of the Indians. In future years his research material would be useful as he cared for and treated the tribal people of Ecuador's rain forest.

Meanwhile, Char Swanson learned that offering hospitality to visitors had always been an important aspect of life in Shell Mera. From the day that Nate and Marj Saint settled in the town, Marj became the only available hostess for miles around. Jungle missionaries on their trips to and from Quito stayed overnight in the Saint home. The alternative was the Esmeraldita Hotel, where everyone slept in the same room, sometimes three to a bed. One night the Saints managed to accommodate 19 guests, among them foreign missionaries, national workers and a schoolteacher and his family. When Mission Aviation Fellowship gave the Saints permission to enlarge their house, Nate increased the number of bedrooms from three to 10.

He also built a large porch at the back of the house, where he put up a clothesline.

Shell Mera has two seasons: rainy and rainier. The downpours often come unannounced, followed by periods of tropical sunshine. Marj Saint, an organized person with a busy schedule, would strip the beds as soon as her guests left. One afternoon the bus going to Quito returned with the same guests that had boarded it a few hours earlier; a landslide had closed the road. Meanwhile, it had rained, and Marj's wash still hung wet on the lines. That night her guests slept on slightly damp bedding.

After Nate's death, Hobey and Olivia Lowrance continued the work with MAF that the Saints had begun. Like Marj, Olivia sorted the mail for jungle missionaries, received their grocery orders, made the purchases and calculated the cost to each. She never knew how many people would descend on her for a meal until Hobey had taxied in from the airstrip. When his passengers deplaned, she counted heads. Experience soon taught her to keep her cupboards well stocked. During the Lowrances' two-and-a-half years in Shell Mera, they hosted 508 overnight guests, most of whom were jungle missionaries.

Several years later, when Gordon and Dorothy Wolfram lived in Shell Mera and the Swansons were once again in Quito, Char brought a group of visitors to the *oriente.* They had planned to return the next day by air, but their flight was canceled. After a long wait at the airport, Char returned to the big house with the news that she and her visitors needed another night's lodging. By then Dorothy had stripped the beds and replaced the used linen with clean bedding, since HCJB had scheduled a tour for the next day. In her inventory of guesthouse linens (a separate supply of bedding marked "guesthouse"), Dorothy had 30 sheets. Her only laundry equipment was a wringer washing machine, and so she had to try to air-dry the laundry between frequent showers of rain. When tours were booked back-to-back, as they were in this instance, she lacked the time needed to wash and dry laundry.

"We'll have to use the same sheets and towels as the visitors used last night," Dorothy informed Char, who well understood her hostess' predicament. Together the two women stripped the beds of the clean sheets, folded them neatly and replaced them with the same linen Dorothy had removed. Such experiences prompted Shell Mera's hostesses never to change guestroom beds until they were certain that their visitors had actually left.

Keeping up with guesthouse laundry was only one concern that hostesses faced. Shell Mera was infested with hardy rats that defied traps. The third floor and the tin roof of the big house became a runway at night as cats chased rats, disturbing the human occupants.

"We *have* to get rid of these rats," Dorothy said to her husband one day. Since the only way to eliminate them was to set out poison, Gordon placed rat poison in the attic, hoping to rid the house of the rodents before the next tour arrived. Unfortunately, double partitions provided hideouts for undetected rats and bats. After eating the poison, one rodent wedged itself into an inaccessible spot behind the bathroom pipes and died there. The smell of the rotting rat lasted for days, and the Wolframs, unable to remove the decaying animal, had to endure the stench that steadily grew worse.

Deodorizing sprays were not readily available in the local shops, and Dorothy had only one precious can that she used sparingly. The day the tour arrived, however, Dorothy slipped into the bathroom with her can and sprayed generously before each guest used the washroom facilities. The mixture of smells was overpowering. That afternoon, as the tour group was leaving for Puyo, one lady handed Dorothy a can of air freshener with the words, "I always like to leave the missionaries something practical."

Although she now had another can of air freshener, Dorothy's problem with rats continued. According to the schedule, the next tour was to stop at her house for afternoon tea. When she saw the group arriving from Puyo, Dorothy went to check her living room. There, to her horror, was a huge rat, obviously poisoned, crawling across her white wool rug. Picking up a metal garbage container, she popped it over the rat and ran for her husband. While Gordon disposed of the sick animal, Dorothy quickly changed the tour schedule, arranging for her guests to visit the hospital before entering her home for refreshments.

Throughout the years the number of visitors to Shell Mera has not diminished. Yet despite the lack of modern conveniences, few hostesses ever complained, for it was visitors who brought them fresh news from the outside world and kept them from feeling completely isolated.

14
The Interest of Others

"Each of you should look not only to your own interests, but also to the interests of others."
– Philippians 2:4

Shell Mera's first Sunday school and church began in 1948 in Nate and Marj Saint's living room. Later, when Gospel Missionary Union moved the Berean Bible Institute to Shell Mera, the Saints' small congregation joined the students at the institute for Sunday worship. At other times, missionaries branched out into the surrounding villages where they held Bible classes for both children and adults. On clinic days, students from the institute distributed literature and shared their Christian faith with people waiting to see the doctor. When Gustavo Arce graduated from the Berean Bible Institute, Epp Memorial Hospital hired him to serve as its chaplain.

As the community of evangelical believers grew, Gospel Missionary Union took the lead in building a concrete-block building across the street from the Bible institute. They dedicated the new church in February 1961, naming it Luz del Evangelio (Gospel Light).. This building housed Shell Mera's first Protestant church. In an area of frame buildings, the church's concrete-block structure seemed elegant, yet its interior was simple. It had wooden floors and was furnished with hard-backed benches and a plain pulpit behind which hung a lopsided picture frame enclosing the words: "For it is by grace you have been saved, through faith--and this not from yourselves; it is the gift of God" (Ephesians 2:8).

Sunday services at Luz del Evangelio were hardly dull. They seldom began on time, and there were usually disturbances, for mangy dogs accompanied their owners to church where they took cover beneath the benches and then sniffed and snarled at one another until a dogfight erupted. One Sunday a young woman visitor was invited to play a clarinet solo during the morning service. She was in the midst of her performance when a stray dog wandered into the church, walked up the aisle, sat down at her feet, and howled "Oooooooo!" non-stop until she finished playing.

Even when the animals were not making a commotion, there was plenty of unscheduled activity. Mothers silenced their crying infants by breast-feeding them while toddlers crawled restlessly along the pews, crunching on crackers. Few adults even noticed this commonplace behavior. When ushers passed the offering plate, some people held it long enough to make change. Near the end of the service, when the preacher's voice usually reached its greatest volume, a bored donkey tied to a nearby tree would emphasize his most telling point with a loud "Hee-haw!"

In spite of such distractions during those early years, church attendance in Shell Mera increased, and members extended the building to accommodate their larger numbers. Throughout the years the little church has been the center of social life for the community's believers, among them a number of Ecuadorian nurse's aides. From this congregation, young men and women have gone abroad to further education at seminaries and Bible colleges and have returned to take up responsible positions in Ecuador's national church and with mission organizations.

Since most of the nurse's aides came from out of town, Epp Memorial Hospital provided them with accommodation on its second floor. Parents of the young girls implored the missionaries not only to provide living quarters for their daughters, but also to be responsible for their conduct. The foreign staff took their responsibilities seriously, for if any one of the girls got into difficulties, her parents would hold the staff accountable. The military base in town increased this obligation since soldiers often wandered along the street, trying to attract the attention of the young women.

One night Art Johnston and Wally Swanson were in the hospital's operating room when a face appeared at the window. Johnston removed his sterile gloves and left the building to investigate. Confronted by the doctor, the prowler, a soldier who had had too much to drink, realized his error and apologized at once. "I'm sorry, doctor," he said. "I didn't mean to disturb you. I was just looking for the door to the *señoritas'* quarters." Art, privately giving thanks that the soldier had met him instead of one of the *señoritas*, accepted the man's apology and steered him back to the main road.

This incident and others alerted the missionaries to the need to tighten their supervision of the nurse's aides, for Ecuadorian culture dictated that the young women must be rigorously protected from the designs of unscrupulous men. Staff posted written rules in a prominent place and made every effort to enforce them. The rules were strict by today's standards--for instance, all *señoritas* were to be in their rooms when the hospital generator was turned off at 9 p.m.--but the behavior of all staff affected the hospital's witness in the community.

In April 1961, Irenne Molina, an Ecuadorian nurse and third-generation Christian, began work at Epp Memorial Hospital. Born and raised in the mountain city of Ambato, Ecuador's fruit and flower center, Irenne had won a Red Cross scholarship to study nursing at Quito's Central University. Wishing to work in a Christian hospital, Irenne, accompanied by her mother, went to Quito after her graduation and applied for work at Rimmer Memorial Hospital. The Quito hospital regretted that it had no openings, but the nursing director told Irenne that one of the nurses in Shell Mera was sick with hepatitis. Would Irenne be willing to work there instead of in Quito?

Irenne agreed to try the Shell Mera position for three months. She would look on it as a three-month vacation, she promised herself, and after that there might be an opening for her in Quito. Accompanied by her father, Irenne traveled to the jungle's edge to take up her temporary post.

Since the accommodation on Epp hospital's second floor was for nurse's aides, Administrator Dave Findlay assigned Irenne a room in the Swansons' big brown house. In order to give the Swanson family some privacy, Irenne ate her meals at a restaurant in town. This gave her ample opportunity to meet the young men of the village, including those stationed at the military base. The men were attracted to this good-looking young woman with shiny black hair, brown eyes and a flawless complexion.

Though Irenne was not her direct responsibility, Char Swanson was concerned about the young nurse who lived in her home. Dedicated to helping others in their Christian walk, Char took Irenne aside and counseled and prayed with her. It was a turning point in Irenne's life. From then on she began to grow spiritually.

Irenne's months at Epp Memorial Hospital passed quickly. Before a nursing position became available in the capital city, Irenne had met Wilson Velasteguí, a young man from Shell Mera and a country boy at heart. Although Wilson lacked the refinement of a city gentleman, he was a Christian. But when Irenne took him to Ambato to meet her parents, *Señor* Molina was upset. He had planned to have his daughter marry a young man in Ambato. But Irenne's mother, Blanca Molina, a devout Christian, knew that the young man her husband favored was not a believer. When Irenne and Wilson returned to Shell Mera, *Señora* Blanca wrote to her daughter, "You may marry Wilson if he is a Christian, but don't bring him here again until you are

Thirty miles south of Shell, Mount Sangay, an active volcano, rises 17,000 feet above sea level.

Farther along the eastern range, Mount El Altar's jagged snow-capped peak can be seen from Shell on clear days.

Above: The original two-room clinic, opened in March 1955, consisted of an examining room, and an obstetrical and operating room. While patients wait to see the doctor, a student from the Berean Bible Institute shares his faith with them.

Right page top: Nate Saint, standing in front of the partially built hospital, never lived to see its completion.

Right page bottom: Epp Memorial Hospital, a 14-bed, frame building at the edge of Ecuador's eastern jungle, opened its doors to patients on May 10, 1958.

Above: Behind Epp Memorial Hospital (left) and at the foot of the water tower is a cement block building that served as workshop and laundry room with apartments for single nurses on its second floor.

Left page top: The Shell-Baños road follows the course of the Pastaza River, winding around steep cliffs with sheer drop-offs.

Left page bottom: The Agoyán Falls, near Baños, changed shape every year from constant erosion.

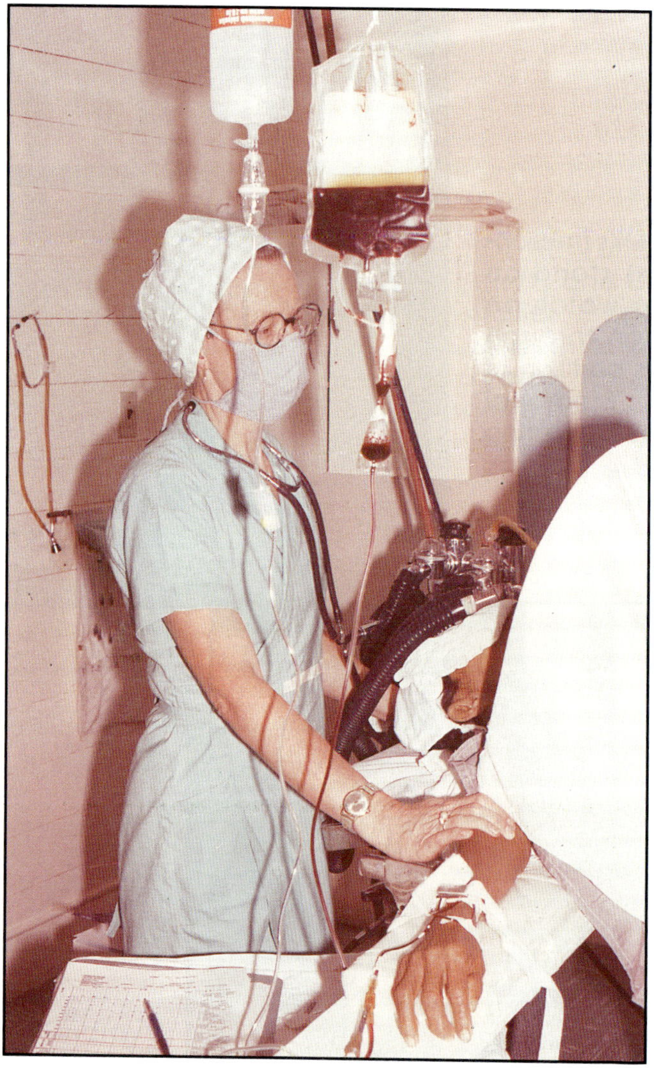

Beth Huddleston administers anesthesia to a surgical patient who, at the same time, is receiving a unit of Beth's blood.

Eleanor Boyes descends the notched-pole "stairs" of a jungle house to help with a clinic that will take place beneath the raised floor.

Joaquin Noroña, pastor of Luz del Evangelio Church in Shell, discusses a Scripture passage with visitors to a rural mobile clinic.

Epp Memorial Hospital (left) stands on the north side of Shell's main road. Opposite, behind the houses, the airstrip runs parallel to the road. Beyond are the military barracks and the homes of afficers.

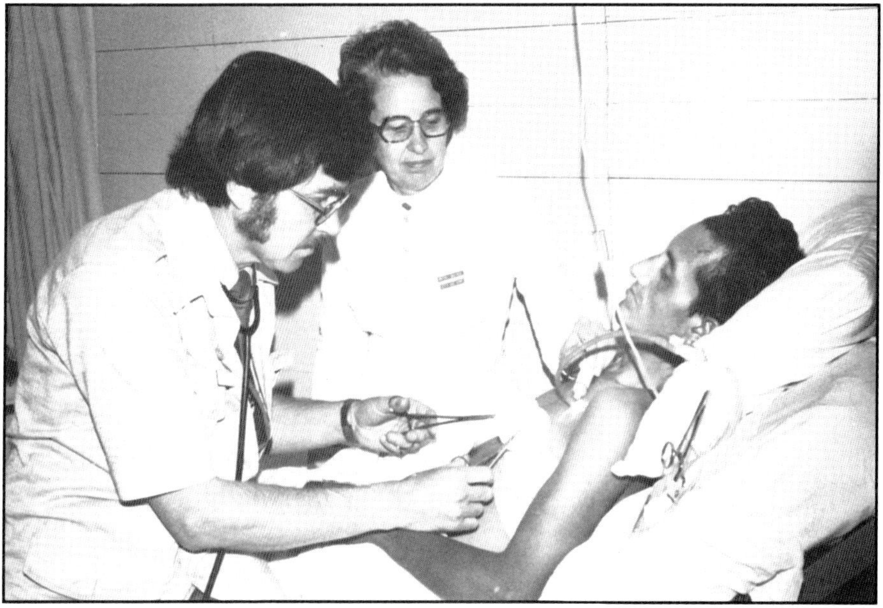

Dr. Wally Swanson and Eleanor Boyes care for an accident victim.

Dr. Ev Fuller writes orders on a patient's chart at the nurses' station, the hub of the hospital's activity.

Señora Rosa smiles happily to be at home with family members while her newborn infant sleeps in a home-made hammock. (See Chapter 22.)

Hospital Chaplain Jesús Montero shares a portion of Scripture with a visitor.

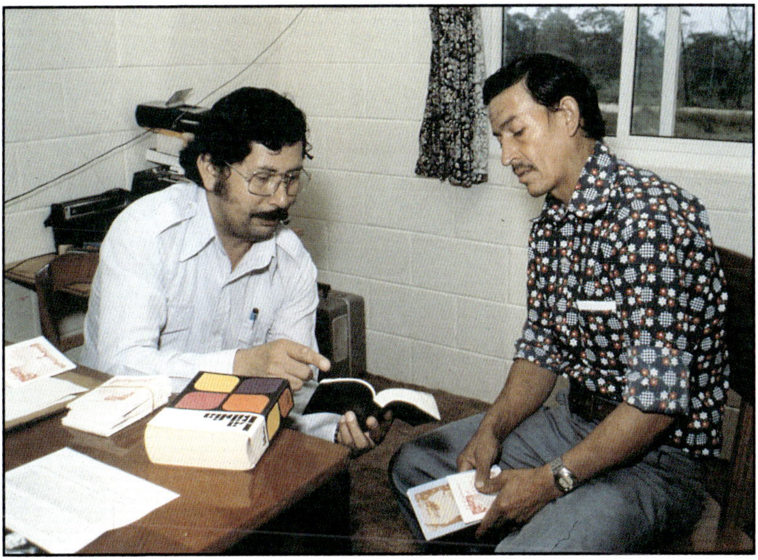

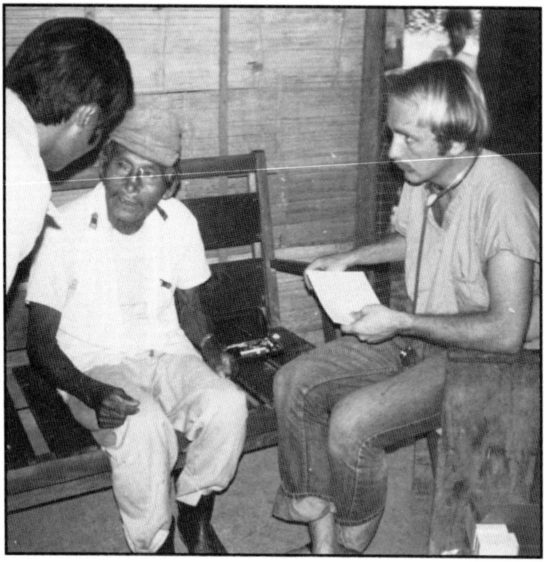

Above: (Left to right) Nurse Anesthetist Dan Swanson, Medical Director Dr. Steve Nelson, Swedish architect Kyell Torbjornsson, Operating-Room Supervisor Beth Huddleston, Hospital Administrator Eleanor Boyes, and project director Ken Edgar formed the team that finalized the designs for the new hospital.

Left page top: At a rural mobile clinic, Dr. Steve Nelson (right), with the help of an interpreter (left), records the health problems of a patient.

Left page bottom: Nurse Judy Cole pauses on the suspension bridge that links the two mission complexes.

Hospital Vozandes-Shell, opened in March 1985, continues to serve the people of Ecuador's rain forest and the surrounding area.

married." Taking her mother's advice, Irenne planned her wedding to take place at Luz del Evangelio Church on April 5, 1963, her birthday.

Wilson, a sports enthusiast, played on the local volleyball team. When his athletic friends found out that he and Irenne were to be married, they decided to attend the wedding. Hoping to divert attention from the ceremony, the people of Shell Mera who opposed the evangelical faith arranged a religious parade to coincide with the hour of the wedding. Late in the day, however, Irenne's parents arrived from Ambato, and her Aunt Inés began at once to prepare food for the guests, thus delaying the ceremony, and so the wedding, scheduled to begin at 7:30 p.m., did not take place until 9 o'clock. By then, the parade had ended, and everyone in the town felt free to attend the wedding. As the bridal party entered the church, the whole village, including the ball team, squeezed into the small building to witness the ceremony.

The newlyweds moved in with the groom's parents, occupying three rooms on the second floor of their home. Irenne kept the little apartment neat and tidy, but the couple had no indoor plumbing and no electricity. Wilson and Irenne would wait nine long years before they could afford to build their own home, and 10 years before their first child was born. When they could finally manage to finance a house, they built across the street from the hospital, naming their new home *Villa Ebenezer*: "Hitherto hath the Lord helped us" (1 Samuel 7:12, KJV).

In the intervening years, Wilson had developed a happy relationship with his father-in-law. When the couple invited their friends to celebrate a housewarming, *Señor* Molina, beaming with pride, carried into the couple's new home many of the delicacies that were to be served to the guests. At the gathering Irenne and Wilson expressed their gratitude to God for His provision of a new home, and Hospital Administrator Norman Emery offered a prayer of dedication.

Irenne's original three-month vacation has extended to more than 30 years. A loyal employee, she has worked with several foreign administrators and a changing missionary staff. The first change came when the Findlays left Shell Mera so that Dave could work on the construction of HCJB's hydroelectric plant at Papallacta.

To replace Dave, HCJB assigned Joe and Alice (La La) Christopher to Shell Mera. The Christophers, who had lived and worked in Panama for several years and still enjoyed good health in retirement, joined HCJB in 1956. Because they were a middle-aged couple, the Epp staff referred to them as *Tío* Pepe and *Tía* La La (Uncle Joe and Aunt La La). Both La La and Joe, accustomed to city life, were appalled at the many inconveniences that the Shell Mera staff took for granted. They proceeded at once to brighten the surroundings of Epp Memorial Hospital in order to bring more pleasure into the life of staff and patients. La La, who loved flowers, instructed the gardener, Agustín, to dig up plants and shrubs native to the area and transplant them to the hospital grounds. Soon yellow hibiscus brightened the border of the parking lot; bougainvillea, a blaze of magenta, crept up the railing leading to the second-floor apartment, and at the front of the hospital, a hedge of pink hibiscus provided the inpatients with a measure of privacy. Behind the building, Agustín planted bushes of perennial orchids and scarlet poinsettias. In keeping with the property's upgraded appearance, Uncle Joe kept the grass cut and the shrubs trimmed. Never before had the hospital grounds looked so attractive.

Providing the hospital with all the electric power it needed was another problem. The town of Shell Mera supplied the hospital's electricity, but its diesel generator was unreliable. The Christophers found an agency in Panama's Canal Zone that was willing to donate a used generator to Epp Memorial Hospital. Though the staff appreciated this good-will gesture, the machine turned out to be very temperamental. It became a constant worry to Uncle Joe, who slept with one ear tuned to its drone. The moment the generator coughed or spluttered and then groaned to a stop, he was out of bed and on his way to the electrical plant. Because the generator was so unpredictable, missionaries kept candles in every room of their houses, ready to be lit the moment the electricity went off. Staff became so accustomed to the generator's hum that the plant's silence wakened them.

Eventually the generator died completely, and Uncle Joe shipped it back to Panama for repairs. For the next nine months the hospital managed with a small plant that could provide electricity either to the missionary homes or to the hospital, but not to all of them. Staff decided that after 7 p.m., the houses would do without electricity. They planned to use candles for the rest of the evening, but they soon discovered that by 8 o'clock they were so sleepy from reading by candlelight that they simply went to bed.

During the Christophers' term at Shell Mera, Uncle Joe supervised the construction of a concrete-block building behind the hospital. The building's ground floor would house a noisy machine shop and, later, a hospital laundry. On the second floor were three apartments for single staff members--not the quietest place to accommodate nurses who would eventually work a night shift. Even so, it was an improvement from the hospital's second floor where the sound of crying babies, the shrieks of women in labor, and the noise of staff rushing to and fro easily penetrated the cracks in the wooden floor boards.

To that point in its history, Epp Memorial Hospital had never been able to provide 24-hour nursing care. To help with emergency operations and to care for the critically ill, nurses would return to work during their off-duty hours. To relieve this situation, the hospital hired Amelia, a nurse's aide, who worked the midnight shift for 10 nights straight. On her nights off, one of the day staff replaced her. The regular staff, unaccustomed to night duty, considered this arrangement a drastic change, for they found it difficult to sleep during the day, and many feared being on duty alone at night. Yet each nurse was expected to take her turn on the unpopular shift.

Olga was already nervous when her turn came to relieve Amelia at night. Unfortunately, she had just gone on duty when she saw a man snooping around the hospital grounds. Staff who had not yet gone to bed came at her summons and chased the prowler away. Then they left, and Olga was alone again. She was just beginning to feel comfortable with her new duty when the doorbell rang. The buzzer's shrill sound sent chills through her body, for she was sure that the prowler had returned. Too frightened to open the door, she ran upstairs to the nurses' quarters, woke her friend Adriana, and then promptly fainted.

Adriana heard the doorbell pealing and knew that someone must need immediate attention. Leaving Olga lying on the floor, she ran downstairs and hurried to the door. At the emergency entrance stood Pablo Freire and his wife Marieta, who was in the last stages of labor. Adriana helped Marieta into the emergency room and then sent Pablo across the road to waken Dr. Johnston. A few minutes later Art hurried into the emergency room to find Adriana, wearing a nightgown and with her hair done up in rollers, holding a crying newborn by his feet. Art took the baby from Adriana and finished the delivery.

Returning to the nurses' quarters, Adriana found Olga recovered from her faint, but too nervous to return to work. Good-naturedly, she removed the rollers from her hair, donned her uniform and finished the shift for her friend. It took time, but little by little, staff adjusted to working the midnight shift.

Dr. Murray Weaver, a surgeon, joined the staff of Epp Memorial Hospital in June 1964. Once a full-time surgeon had been added to the staff, the number of surgical cases increased steadily and it became necessary for the hospital to expand. Thus, in 1965, workmen added a new, concrete-block, outpatient department that extended from the emergency room to within a few feet of Shell Mera's main road. The new addition housed doctors' examining rooms, the X-ray department, the pharmacy and the laboratory. The addition made the hospital into an L-shaped structure and freed rooms for additional inpatients, increasing its capacity from 14 beds to 25. The original two-room clinic now became an isolation unit for patients with contagious diseases.

When Zulay Cárdenas, a graduate of the Berean Bible Institute, returned from Honduras as a qualified laboratory technician, the hospital hired her. Since she was not needed in the lab on a full-time basis, the doctors taught her to take X-rays and to fill their prescriptions from the stock of medicines in the pharmacy. Zulay learned these additional duties well and for several years ran the three departments. Later, when the number of people coming to the outpatient department increased, staff released Adriana from the nursing schedule, and she became the hospital's pharmacist. In time, both women became long-time employees, participating in all of the hospital's activities, on and off duty. Zulay has since died, but it can truly be said that these long-time staff members were key people in establishing the family spirit that existed within Epp Memorial Hospital.

15
Disasters

"Do not boast about tomorrow, for you do not know what a day may bring forth."

– Proverbs 27:1

"HCP31, Shell Mera! Shell Mera! HCP34, Quito calling Shell Mera!" The voice coming over the airwaves was urgent. La La Christopher, standing in the hospital corridor opposite the tiny radio room, heard Epp's familiar call letters and responded immediately. Picking up the microphone, she spoke into it, "HCP31 answering. Go ahead, Quito."

Speaking from his Quito hangar, Gonzalo Ruales of TAO (*Transportes Aéreos Orientales*) answered, "Captain Díaz (a young Ecuadorian pilot) is doing my regular flight between Macas and Shell Mera today. I've been on the radio with him ever since he took off from Macas, but his communications suddenly stopped." He paused and then, with a note of fear in his voice, added, "He should have landed in Shell Mera five minutes ago. I'm afraid his plane has gone down!"

Having flown the Macas-Shell Mera route many times, Gonzalo knew exactly where the missing aircraft should be. He wanted *Tío* Pepe to search for his plane. La La scribbled the information Gonzalo gave her on a notepad and assured him that she would call him back as soon as she had news.

With the scribbled information, portable stretchers, and members of the hospital staff, *Tío* Pepe drove up the road leading from the hospital toward Mera. His search was brief, for within a few minutes he spotted Gonzalo's plane lying twisted and crumpled on a small knoll not far from the main road. The wreckage was precisely where Gonzalo had predicted that it would be. A light drizzle was falling, and thick overhanging clouds that hid the town from aerial view had caused the young pilot to overshoot Shell Mera's landing field.

With heavy hearts, Uncle Joe and his companions hurried toward the smoldering wreckage. The townspeople, who had witnessed the plane's descent, had arrived on foot and were already searching through the charred ruins for the bodies of Captain Díaz and his three passengers. There were no survivors. Among the wreckage Uncle Joe found the body of Salomón Balseca, Gonzalo's cabin boy and the teenage brother of the Christophers' maid, Bertha. It appeared that the lad had crawled from the burning plane on his hands and knees, for he was several feet from the wreckage, and his arms and legs were burned. While the townspeople extricated the bodies of the three passengers and Captain Díaz from the downed plane, Uncle Joe and other members of the hospital staff carefully wrapped the body of Salomón Balseca in canvas.

Back at the Christophers' home, Joe and La La laid out Salomón's body on their coffee table in preparation for the wake. Local custom demanded that the body of the deceased be "accompanied" by grieving family members and solemn-faced neighbors through an all-night vigil. That night the townspeople came to express their sympathy and to join the night watch. Late in the evening, La La served the mourners and their friends the traditional cookies and cinnamon tea to fortify them for the rest of the night. At most wakes in the area, the tea is heavily laced with *pecho amarillo*, a cheap liquor made locally, but La La's tea contained only cinnamon.

At midnight, *Señora* Balseca, the cabin boy's mother, announced that everyone should go home. As a Christian she considered it pointless to sit up all night with a body from whom the soul had already departed to be with the Lord. Although her suggestion was contrary to local custom, out of respect for the elderly woman, the mourners followed her advice and left.

That night Gonzalo Ruales traveled overland from Quito, arriving in Shell Mera the next morning with two well-made caskets, one for the body of Captain Díaz and the other for Salomón. The bodies of the three passengers had already been flown home to Macas. When the curious townspeople saw the casket, they lined the veranda of the Christophers' home, pressing their noses against the screened windows, hoping for a glimpse of the charred body. Someone managed to shoo them away while the next of kin placed Salomón's canvas-wrapped body into the coffin and nailed down the lid. The Balseca family was grateful for Gonzalo's gift, for only the wealthy could afford a ready-made casket. After a brief burial service, male relatives carried Salomón's casket on their shoulders to its final resting place. A sad reminder of the preceding evening was that the steam from Salomón's body and its damp canvas cover had turned the top of the Christophers' coffee table white.

In Ecuador the immediate family takes responsibility for all the arrangements following the death of a relative. Thus no one at Epp Memorial Hospital was surprised when the Huanga family from Madre Tierra (Mother Earth) drove into the hospital yard one day in a makeshift passenger carrier that contained the rigid body of their mother. She had died at home that morning after a lengthy illness. Because the Huangas could not afford a ready-made casket, they knew that they must either wrap their beloved mother in a blanket or fashion a rough box from whatever wood they could manage to obtain. Faced with these unattractive alternatives, the whole family had piled into the carrier beside the corpse and, weeping openly, had come to ask Angel, the hospital carpenter, to make a casket.

This was not the first coffin Angel had made, nor would it

be the last. As he usually did on such occasions, he took his ruler and measured the length and width of the dead woman's body in order to calculate the amount of lumber he would need to build a rough box; then he went off to buy the materials. Never one to move quickly, he took several hours to make the simple coffin.

Meanwhile the open carrier stood in the hospital's driveway, in full view of visitors to the outpatient clinic while the tropical sun accelerated the rapid decomposition of *Señora* Huanga's body. Unable to hurry the carpenter and wishing to give a measure of dignity to the unpleasant scene, La La carefully covered the dead woman. When Angel finally finished making the casket, the grieving family left for home to prepare for the customary wake.

Although for lengthy periods of time life in the *oriente* drifted by without major incident, many of life's dramas involved the Shell Mera staff, who learned to contend with situations seldom encountered in the homeland. One dreadful tragedy unfolded in 1967 just as the nurses were finishing the day shift. It had begun innocently enough on a Saturday afternoon when Rafael Yépez, an airplane mechanic, took his family for a picnic and a swim at the Alpayacu River. After spending the best part of the day in the sun and water, Rafael, his wife and teenage daughter, Rocío, climbed into the cab of the family's small panel truck. The younger children, including their guest, the son of an Ecuadorian army major, crawled into the back of the truck. As they headed for home, the sunny skies clouded over, and a heavy downpour began.

To reach their home at the army base, Rafael had to drive across the airstrip. Since the control operator had left the observation tower after the last plane took off, no siren warned Rafael that an aircraft was approaching at the far end of the runway. Failing to check the airstrip for activity, he started across in the pouring rain.

Overhead, dark clouds obscured the eastern range of the Andes Mountains, and sheets of rain sprayed the Plexiglas of the DC-3 as it descended onto Shell Mera's runway. Moments before, Captain Pico, a retired Ecuadorian Air Force colonel and experienced pilot, had taken off for Quito, but poor visibility had forced him to return. With eyes focused on the airstrip below, he was quick to notice the truck in his path. He had slowed the plane for landing, but now he accelerated for a pull-up. Passengers on the aircraft had braced themselves for a bumpy landing on the gravel airfield, but instead their plane suddenly shot upwards. Once the DC-3 was airborne again, Captain Pico took a deep breath in relief. He thought that he had averted a crash with the truck and saved the lives of his passengers.

Captain Pico circled the town and, this time, brought the aircraft to a safe landing on Shell Mera's runway. But unknown to him, during the plane's first descent, its landing gear had struck the top of Rafael's truck, slicing off the cabin roof and fatally injuring Rafael. His pregnant wife, sitting next to him, suffered only minor injuries. Rocío, who was sitting forward on the front seat, received serious facial injuries when a jagged piece of metal sliced through one eye, across the bridge of her nose, and through the other eye, permanently blinding her. When they heard the thunder of the plane overhead, the children in the rear threw themselves on the floor, escaping injury.

Within minutes, nurses at Epp Hospital, just going off day duty, were startled by the sound of people running down the corridor pushing stretchers that held the accident victims. Soon relatives and friends of the injured swarmed into the hospital. The town was in an uproar. Family members, hysterical and angry, refused to listen to reason. One relative had died, and another would be handicapped for life. Some talked of retaliation. To avoid a confrontation with the Yépez family, the innocent pilot went into hiding for several months. Finally, talk of the incident subsided, and life, for a time, returned to normal.

One evening, however, a truck loaded with gasoline drums drove into Shell Mera and parked in the center of town where a number of people were milling about. The truck's driver, Alfonso Pazmiño, noticing a strong odor of gasoline, tried to locate the leak. To help him, someone lit a candle, but when Alfonso opened one of the drums, gasoline poured onto the ground, forming a pool that instantly ignited. Soon the open drum was ablaze.

Quick to realize the danger of fire in a town of wooden buildings, Lucho Guerrero, a courageous bystander, climbed into the truck's cab and drove it away from the town. As he drove, the burning truck became a roaring inferno. Onlookers watched aghast as it sailed past the hospital, the MAF hangar, and around the curve beyond the end of the runway. There it veered off the road into an open field owned by the tea company. As it left the road, Lucho jumped clear of the cab, only moments before the truck exploded. Drum after drum of gasoline burst into flames. The explosions lit the tropical sky, and the air vibrated with the detonations.

Thinking that he could cover his losses by going to court, the truck's owner tried to sue Lucho, accusing him of stealing his truck and then damaging it. To the townspeople, Lucho had performed an heroic act that had saved Shell Mera from destruction. They supported him by signing a deposition that cleared him of all damages.

It was late on a Sunday morning of September 1964 when a chartered military plane belonging to *Fuerza Aérea Ecuatoriana* (Ecuadorian Air Force, or FAE) loaded for take-off at *Río Amazonas*, Shell Mera's airport. On board the plane were 20 doctors, Ecuadorian and foreign; the latter were pediatricians from Ship Hope, a mercy vessel anchored in Ecuador's port city of Guayaquil.

Project Hope (Health Opportunity for People Everywhere) had recruited top-notch doctors and nurses for its mission of mercy. Its medical team lectured, visited local hospitals, performed major operations and demonstrated modern medical techniques. During the ship's stay in Ecuador, the medical personnel aboard had inoculated a half million people against diphtheria, polio, tetanus and whooping cough. After a busy 10 months, the ship and its passengers would soon return to the United States.

On this outing, Ship Hope's physicians were guests of a number of Quito doctors. On Saturday, they had visited military and government medical services in the area, as well as flying to Limoncocha, Wycliffe Bible Translators' impressive base camp deep in the Amazon rain forest. Now, on this bright sunny morning, the doctors boarded the aircraft for their return trip to Quito and Guayaquil. One of the passengers, Dr. Anne O'Neil* of Ireland, Ship Hope's head pediatrician, tried to fasten her seat belt, but found that the strap was broken. At that moment, Captain Morales, FAE's pilot, revved the plane's motor and roared down the airstrip. As the aircraft lifted from the ground, one engine failed. The nervous pilot circled to land the crippled plane, but as he lost height, he was unable to line up with the runway. When he tried to pull away to repeat his approach, the plane failed to gain the necessary altitude. In panic, Captain Morales banked sharply to the left, and in the next instant the plane's left wing clipped a tall tree. The plane crashed into the forest near the Pastaza River.

At Mission Aviation Fellowship, Carol Osterhus saw the plane go down. She alerted her husband, Dave, who rushed to the church to inform Dr. Wally Swanson. The Christophers, who were also in church, heard the plane take off and knew that the pilot must be in trouble because he was flying dangerously low. They, too, hurried to the hospital to prepare for an emergency.

Warren Blanchard and Lois Price, newly arrived nurses at Epp Memorial Hospital, were returning from church in Mera. As they drove into town, they were puzzled to see the people of

---

* Name has been changed.

Shell Mera standing by the side of the road gazing into the sky. When they reached the hospital, Uncle Joe told them of the plane crash. As Lois and Warren prepared to join the rescue team, Uncle Joe put his hand on Lois' shoulder and said, "Honey, you'd better put on your uniform so that the local people will know you're a nurse."

Taking his advice, Lois hurried to her apartment, where she quickly changed into a clean white uniform. Meanwhile, MAF pilot Dave Osterhus, flying overhead, circled continuously but saw no sign of the downed plane. It was not until he checked an area where the trees appeared disturbed that he saw the damaged plane, which was lying midway between the airstrip and the Pastaza River. On impact, the treetops had parted and then closed in again, concealing the wrecked aircraft. The uninjured survivors, who had been thrown in every direction, were walking away from the wreckage.

Warren and Lois set out with Dr. Wally Swanson in the hospital's Land Rover. At the southeast end of the runway, Wally took the dirt road that leads to the right, heading toward the village of Madre Tierra and the Pastaza River. Parking the jeep by the side of the road, he and the two nurses struggled through thick underbrush and shrubbery until they came to an embankment. Lois, dressed in her white uniform, followed Wally and Warren, sliding down the muddy slope and covering herself in mud. Later, the townspeople mistook her for one of the accident victims.

Wally, equipped with a walkie-talkie, kept in touch with Dave Osterhus, who directed him to the downed aircraft. Soon they found Dr. Arcos, a Quito physician, covered in mud and in severe pain from multiple rib fractures. They carried him out to the parked Land Rover and made him as comfortable as possible. Then Wally returned to the crash site in search of Dr. Nicholas Espinosa, a pediatrician from Rimmer Memorial Hospital in Quito. Nicholas, who had visited Wally the day before, was nowhere to be found. Wally wanted to search the aircraft, but military authorities prevented him from entering the plane moments before it burst into flames. When Wally finally found Nicholas, he was on another path, limping along on a fractured leg.

Back at Epp Memorial Hospital, staff hastily admitted the 20 injured doctors. Since the hospital had only 14 beds, the nurses placed mattresses on the floor in rooms and along the corridor. The hospital was bedlam. One disoriented surgeon wanted to get up and operate. The doctors' brush with death, which changed their role from that of visiting physicians to patients, had left them jittery. The most critical patient was the lovely Anne O'Neil. As Wally Swanson stitched the cuts in her face, she repeatedly asked, "Am I going to die?" That night La La Christopher slept on the floor in the hospital's linen closet. Though she had no medical training, she wanted to be available to help the nurses by caring for patients with minor injuries.

The crash victims did not remain at Epp Hospital for long. Ship Hope's chief of staff considered that his well-qualified medical personnel could care adequately for the injured on the ship's return journey to its home port in the United States. When Dr. O'Neil arrived in Guayaquil, her colleagues immediately arranged to fly her to the States, but it was too late. During the flight her condition worsened, requiring an emergency landing in Jamaica, where she died.

Doctors from Quito came to Shell Mera to arrange flights for their colleagues. One pediatrician, though, remained behind. Dr. Nicholas Espinosa adamantly refused to return to the capital by air. Instead, he waited at Epp Hospital until his brother Fabian, also a Quito physician, could leave his practice and drive to Shell Mera to take him home overland.

The plane crash was a traumatic experience for all concerned. Yet this accident had two positive consequences. Realizing that the four crew members had died in the crash, the visiting physicians considered themselves fortunate to have survived and to have received care in a clean, efficient hospital that they had not known existed. Several of these doctors wrote to Epp Hospital to express their gratitude, and their letters encouraged the staff.

One Ecuadorian physician never forgot the doctor who had rescued him. Later, Wally Swanson took his oral exams at the Central University's Medical School in Quito to validate his medical license in Ecuador. In the absence of the dean, Dr. Arcos, a survivor of the crash at Shell Mera, was in charge. Recognizing Wally, he turned to his Ecuadorian colleagues and said, "This doctor saved my life."

16
The Little Ones

"Let the little children come to me, and do not hinder them, for the kingdom of God belongs to such as these."

– Mark 10:14b

As a young pilot pioneering Mission Aviation Fellowship's work in the *oriente*, Nate Saint had envisioned not only a hospital to serve the people of Ecuador's eastern jungle region, but also a school for missionary children. Such a school would allow parents to remain united with their younger children rather than having them go away, at an early age, to a boarding school. Although Nate had bought property for a school, he did not live to see the fulfillment of his dream. But the missionaries who followed Nate quickly saw the need for an elementary school facility for the MKs (missionary kids) at Shell Mera. One such couple was Dr. Art and Verna Johnston.

While they were on furlough in Canada in 1959, the Johnstons arranged for Estella Primerman, an elementary school teacher, to accompany them back to Ecuador. Estella opened a school in the large house next door to the hospital where she taught the Johnston and Woods children. But ill health forced her to return to Canada at the end of the school year. Next, Miriam Richards, a retired schoolteacher came to teach the missionary children of Shell Mera.

She, too, taught for a year, but a heart attack prevented

her from continuing.

Then Dr. Wally and Char Swanson joined the missionaries at Shell Mera. Tod, their six-year-old son, was ready for school, and Char, who was herself a teacher, began to teach him at home. Later she opened her classes to the children of fellow missionaries, adding more grades until the little school on the second floor of the spacious house had ten or more children on its roster, in addition to a nursery group made up of four or five toddlers. Thus, though the vision of a school originated with Nate, it was the Swansons who brought his dream to fruition.

In preparation for the Swansons' furlough, which was to begin in the summer of 1965, HCJB's Liz Kimmer, a teacher at the Alliance Academy, moved from Quito to Shell Mera to open the school's fall term. Until the missionary community was able to erect a building, Liz continued to teach in the roomy house. She had plenty of adjustments to make since each of her six pupils was in a different grade, and school supplies were scarce. There was, for instance, only one history book available for the sixth grade. Liz arranged for student Mervin Schmidt to use it during school hours, while she took it home at night to prepare the next day's lesson.

* * *

It was a glorious sunny day in August 1965, the beginning of the school term, when missionaries gathered to start erecting a school building at Shell Mera. Fathers and sons mixed and poured the cement for the foundation of the long-awaited building. During the next five months, the new structure took shape under the direction of Ralph Sandall, a builder and carpenter on loan from Wycliffe's jungle base at Limoncocha. Local missionaries helped Ralph to obtain the necessary construction materials, while funds donated for the project were channeled through HCJB.

The new school, built midway between MAF and the hospital, was plain and simple, resembling two large boxes of the same size and shape, placed one on top of the other. Like other buildings constructed by foreigners living in a wet climate, it had wide overhanging eaves. Its wooden walls were unvarnished, its windows were unglazed, and sand was piled high in several heaps around it.

By January 8, 1966, the tenth anniversary of the martyrdom at "Palm Beach," the new school was ready to be dedicated. The weather was forbidding, for Mount Sangay was hidden by rain and fog. MAF's Ted Lepper had whitewashed the top of a cone-shaped sand pile, and to a bus load of ladies who had accompanied Marj Saint from Quito to be present on this historic occasion, he announced, "This is Sangay." The ladies were laughing as they went into the new building for the dedicatory service.

The school's interior was as unimpressive as its exterior. No pictures decorated the unpainted walls; the tile floor, cold and uninviting, had been laid only the night before; boxes of books, still to be unpacked, were stacked in one corner; and on this rainy morning, damp air drifted through the window openings. Only a green chalkboard and a few children's desks identified the room as a school.

Local missionaries and visitors assembled in one of the two classrooms, where chairs had been arranged for the simple opening ceremony. School Board President Dave Osterhus welcomed the guests. Ted and Nancy Lepper sang a duet, and the nine school children sang the hymn, "We Rest on Thee." It was the hymn that the five missionaries had sung on the morning they entered Aucaland.

Nate Saint's sister Rachel, who was then working with the Waorani (Auca) tribe, described their jungle school. Although many of the Waorani children needed an education, Rachel and her co-worker were teaching the adults first. After the men returned from their day's activity of blowgun hunting, the two women taught them at night by the light of a gas lantern.

"If we had a Parent-Teachers' Association," Rachel said, "most of the students would be in the PTA. But we have some very special people in our branch school," she went on. "Four of the five men who killed the missionaries are students. At first, Uncle Gikita, one of the killers, was not interested. Then, when he saw the Gospel of Mark in print and heard it read, he, too, wanted to learn to read."

The opening ceremony ended with HCJB's Bob Savage offering a dedicatory prayer: "Father in heaven, take the cement and the bricks, the window panes and wood, the nails and the paint. Use this building so that the children will not only learn their ABCs, but also how to walk with Thee."

Bob's prayer has been answered. Despite the distractions of rain pounding on the roof, the deafening roar of planes taking off and landing on the runway, soldiers trotting up the road repeating their sergeant's commands at the top of their lungs and the rat-a-tat-tat of military guns used for target practice, students at the Nate Saint Memorial School do learn more than their ABCs. Their school day begins with morning devotions, helping them to develop a close walk with the Lord. With complete trust and faith in God, they bring their petitions to the Heavenly Father who hears and answers their prayers.

One of the children's prayers was for Steve, a foreign medical student, who spent three months at Epp Memorial Hospital. As Steve was preparing to leave for his homeland, he said to the teacher, "Tell the children that God has answered their prayers for me. I now believe in Christ."

The children often pray for patients in the hospital. One of the patients for whom they prayed was Hilda, a happy five-year-old girl running up and down the sidewalk behind the hospital, her black hair parted in the center and plaited in two long braids that bounced as she ran. She pulled a toy wagon on a string that was pinned securely to one of her two empty sleeves. The curious school children asked, "What happened to her? Why doesn't she have arms?"

Hilda was the victim of a *trapiche* (sugar-cane press) accident. One day, while she was feeding sugar-cane stalks into her family's *trapiche*, Hilda's foot slipped, sending her hand into the press along with the sugar cane. Instinctively, Hilda tried to use her other hand to extract the one that was caught in the crusher, and the unforgiving machine shredded both her hands and forearms. MAF flew Hilda from Macas to Shell Mera, where hospital staff rushed her to the operating room. Since both her forearms were gone, the doctors had no choice but to amputate her arms just below the elbows, though in such cases they tried to save as much of the limbs as possible, knowing that most patients are unable to afford a prosthesis. Since Hilda was young and still growing, any artificial limbs would have to be replaced every few years, and the doctors doubted that her family could afford that expense.

In the hospital at the same time as Hilda were two little boys, victims of snakebites. Panchu had been bitten on the leg, but eight-year-old Milton was bitten on the head. He had been playing on the riverbank when a bushmaster, the largest of the world's pit vipers and South America's longest venomous snake, bit him, striking three times. Milton's family rushed him to a clinic in nearby Tena where he received antivenin, but no antibiotics. Eleven days later, with his head swollen to twice its normal size and his eyes mere slits, Milton was flown to Epp Memorial Hospital. Bacteria in the snake's mouth had caused a severe infection.

When the swelling persisted even after a series of antibiotics was administered, the doctors concluded that the infection under the scalp needed to be drained. The procedure could leave Milton with noticeable scars. Since Milton's family was Quichua, the doctors feared that its members might be critical of the treatment, especially if Milton died.

Their concerns were unfounded, for forty years earlier Milton's father, Simón Grefa, had worked for Reuben Larson at Dos Ríos. Simón had proved to be so trustworthy that Reuben would leave his family in the care of this Indian whenever he was away from home. While working for the Larsons, Simón had confessed his faith in Jesus Christ as his Savior. Now, years later, when the doctors explained to him the treatment they proposed, Simón replied, "My son has a right to live. Please do whatever is necessary to save his life."

To receive the specialized treatment he needed, Milton was flown to Rimmer Memorial Hospital in Quito, where doctors made an incision at the base of his skull and inserted a tube. Through this opening, Milton received massive doses of antibiotics. Finally, his grotesquely swollen head returned to normal size, and skin grafts were placed onto the areas of his scalp where gangrene had eaten away skin and flesh.

Simón's faith in the doctors' skills was justified. Milton survived the treatment in Quito and returned to Shell Mera to recuperate. With a white halo of gauze wrapped around his head, Milton was a little jungle charmer, smiling through the ordeal. The staff loved him and were filled with joy as they watched his condition improve. Later he could push Panchu, whose leg was bandaged, in a wheelchair while Hilda ran behind. Together the three children scampered up and down the hospital corridor, laughing and giggling. Milton's comment on his misadventure was, "*Aunque me haya picado, me gusta*" (Even though the snake bit me, I like it)."

17
Adversity

"We were under great pressure, far beyond our ability to endure."

– 2 Corinthians 1:8b

Shell Mera's military base was the largest in the *oriente*. In 1941, before the military moved into the area, the Peruvian Army had invaded Ecuador's southeastern frontier and had claimed an extensive part of the country's Amazon basin. In settling the two countries' counterclaims, the Latin American Council ceded the disputed territory to Peru, but Ecuador had never recognized the new border.

Then, in January 1967, 25 years after the signing of the *Protocolo de Río de Janeiro*, word reached Ecuador that Peru was planning to place cement markers along the frontier. Fearing a Peruvian assault, Ecuador's military force stationed at Shell Mera took the offensive. Jittery army officers sent out an alarm and air-raid practice began. The base commander ordered the civilians of Shell Mera to evacuate their town, placed Epp Memorial Hospital under military control, forbade medical personnel to leave the area, and ordered half the hospital's staff to serve on mobile teams and the other half to remain at the hospital to care for the casualties that he anticipated. To prevent the hospital from becoming a target of aerial attack, he ordered Administrator Ralph Horn to paint a red cross on its roof and on the water tower.

The missionaries at Shell Mera had planned to leave for Quito to attend HCJB's annual members' meeting, but now they quickly prepared for a possible air assault. Ralph Horn and handyman Agustín climbed the water tower and mounted the hospital roof, hurriedly painting a red cross on each. To prevent enemy aircraft from landing, soldiers cluttered the runway with discarded oil drums and other debris. Most of the local people left the area, but before they did so, they brought their valuables to the hospital, where staff labeled them and stored them in the safe.

That night the Christians who remained at Shell Mera met in Luz del Evangelio church to pray, asking the Lord to protect their town. They encouraged one another and found comfort in the words of the Psalmist: "He who dwells in the shelter of the most High will rest in the shadow of the Almighty" (Psalm 91:1).

The military expected Peruvian forces to attack on Sunday, January 29, the anniversary date of the signing of the Protocol. That morning the remaining residents of Shell Mera awoke to steady rain and one of the lowest cloud covers ever reported for the region. Teachers and students at the Berean Bible Institute rose before dawn, showered and dressed. Leaving their dormitories before daylight, they walked down the road toward the hospital. The atmosphere of the vacated town was eerie, almost ominous. Earlier that morning the few remaining inhabitants had slipped down the banks of the wooded ravine to hide under the overhanging branches screening the Motolo River.

At the edge of the wooded ravine to the rear of the hospital, teachers and students sought refuge in the "Auca Hut." This gazebo-like shelter, a favorite place for staff picnics and cookouts, resembled a jungle hut with half-walls marking out its floor space and a roof of thick thatch. It was an unlikely haven during an air raid, but the little group of civilians, having placed their faith in the Lord and not in this primitive shelter, waited there in peace. Except for the rain, all was quiet.

"We really didn't believe that anything would happen," one missionary recalls. "Finally we decided to hold a church service and Sunday school."

The rain continued all day until late afternoon. When it stopped, the sky cleared, and the most glorious sunset appeared in the western sky. By now it was too late in the day for any Peruvian planes to reach Shell Mera. With the anniversary of the protocol past, Christians thanked the Lord for His protection, and life in Shell Mera returned to normal.

Once a jungle outpost for a foreign oil company and now a government military base, Shell Mera steadily grew to a population of more than a thousand people. It had no formal links with the municipality of nearby Mera, and when it had achieved the status of a parish, it dropped its second name and became known simply as Shell. By 1968 a position opened for a *Presidente de la Junta Parroquial* (President of the Parochial Board) of Shell. When public opinion in Shell proved to be polarized, a delegation from the town approached Dr. Art Johnston to ask if he would fill the post until an election could be held. He agreed to do so.

Dr. Johnston, who held Ecuadorian as well as Canadian citizenship, was eligible for the position. Born and raised by missionary parents in northern Ecuador, he carried an Ecuadorian passport and, unlike foreigners, had an identification card that was the same color as those of other Ecuadorians. He was fluent in Spanish, understood the Ecuadorian culture and loved the people. Since he had spent most of his missionary career in Shell, his fame had filtered into every corner of the *oriente*.

After Art filled in pre-term, town officials persuaded him to run in the next election. When they counted the votes, the doctor had won. He welcomed the opportunity to help these people, many of whom were his patients. During the years he had spent in Shell Mera, Art had listened to the complaints of the townspeople, knew their wishes, and wanted to provide some worthwhile improvements in the town. During his term of office, therefore, he worked to bring about some important measures. One was the construction of a public washroom, for the town had no toilets. Since homes lacked indoor plumbing, people had long been accustomed to using the weeds by the side of the road. Art knew very well the health hazards of this practice, for every day he prescribed medication for their infestations of parasites.

Another of Art's achievements was the construction of a stone wall at one corner of the plaza where the townspeople threw their garbage into a ravine. To avoid falling over the steep embankment, people had tossed their waste into the air, and much of it littered the town square. To reduce the amount of litter and to protect the townspeople from accidents, the young doctor had organized a work crew to build the wall.

The townspeople deplored the fact that Shell's electricity was so unreliable. Lights were sometimes unavailable for several nights at a stretch. The military used the old light plant that had belonged to the Shell Oil Company.

Technically, it supplied both the army base and the town with electricity, but the town got only enough for each home to have one naked light bulb dangling from the ceiling. Town officials found a sympathetic ear in a member of the Ecuadorian Congress from Puyo who had local connections in Shell. With the money that this congressman obtained, the town bought a new power plant. Art arranged for the building of a road that ran down the slope beside Luz del Evangelio Church. The men installed the new light plant at the foot of this hill where the steep cliff muffled the sound of the noisy generator.

Dr. Art Johnston held the position of *presidente* for more than a year. Though he continued to serve at the hospital, he also enjoyed his work with the town officials, who appreciated his concern for the welfare of the local people. His community work, a diversion from the practice of medicine, gave Art a sense of fulfillment in helping the Ecuadorians he loved. Yet divided interests can create problems, and in missionary work, it is not always easy to reach a happy balance. In their effort to run an efficient hospital, the nurses often became frustrated when Dr. Johnston could not be found. Eventually he would appear, but not before the nurses' patience had worn thin.

As the town's population expanded, so did the medical work at Epp Memorial Hospital. Soon the washing of the hospital's laundry became an impossible task. For several years the soiled linen had been distributed among three local women who washed it in the river. Later the hospital contracted Victoria Saramiento, a *mestiza*, whose house overlooked the Motolo River at the far end of town. She employed Indians to do the work, and the hospital paid her according to the weight of the linen washed. Three times a week the hospital's dirty laundry was counted, bundled into bags, and carried on the backs of Indian lads to the river's edge. Indian women descended the steep riverbank to scrub the soiled sheets on the rocks in the Motolo River. The women whacked the clothes against the rocks with such force that they quickly wore out. They stretched the clean linen on the grass and over nearby bushes to dry. In between tropical showers, the hot equatorial sun dried and purified the items.

In those formative years, visitors were fascinated by the sight of Indian girls strolling through town on their way to the hospital, each one balancing a basket of clean linen on her head. Yet to the nurses it seemed that the girls always arrived at the hospital at the busiest time of the day. To control the hospital's inventory, the nurses had to interrupt their regular duties to count the clean linen, checking each item against the list of soiled laundry, and then weigh the bags. To increase her pay, Victoria would return some linen to the hospital damp. She left the wettest items on the bushes to dry, which meant that no single day's laundry ever returned to the hospital complete.

To maintain an adequate supply of linen, nurse's aides mended the torn laundry and sewed new items during slack periods. But trying to make do with a limited supply of clean linen was a constant frustration. During rainy periods, it took several days for the clothes to dry. Staff ran out of ideas on how to skimp and stretch the few items that remained in the cupboard. Often the nursery was full of newborns who needed a steady supply of clean diapers, while down the hall in pediatrics, small children suffering from diarrhea and vomiting needed both diapers and clean bedding. Disposable diapers were unheard of and, if obtainable, would have been too expensive for a jungle hospital's limited budget.

Though washing the linen in the river was acceptable by local standards, the hospital staff was concerned about the practice. It contaminated the water that served other settlements downriver, and this hazard, along with the continual lack of clean linen, were factors that demanded setting up a hospital laundry. But to take such an initiative was not as simple as it sounds. By the time that the hospital administration approved an unbudgeted expenditure and did the paperwork required by the government to obtain permission to import equipment, several months had elapsed. Finally, however, when all the red tape had been handled successfully, the Christophers bought two domestic wringer washing machines that for the next several years served to launder the hospital's linen.

La La Christopher offered Victoria the position of hospital laundress, but Victoria refused. She was an astute enough businesswoman to know that if she accepted La La's offer, she and not the Indian women would wash the dirty clothes and bedding. It had suited Victoria very well to hire Indians to do the hard work for a mere pittance while she pocketed most of the money paid by the hospital. The salary that the hospital now offered her as laundress was less than she had received by contracting out the laundry under the old arrangement.

Since Victoria refused the position offered, she suddenly lost her income. Never one to take a loss quietly, she soon found a lawyer who agreed to represent her in bringing a lawsuit against the hospital. In bringing this suit, the lawyer had Victoria name Administrator Gordon Wolfram and Medical Director Art Johnston as codefendants.

The two men appeared at the preliminary court, where an opportunity was offered for reconciliation. Victoria's demands were high: she claimed severance benefits as an employee who had been fired unjustly. Since Gordon and Art could not agree to meet her demands, they were launched on months of legal wrangling for which they had no training. The stresses of handling the suit during the next several months drained them physically and emotionally.

Distance made it difficult for the two men to consult with colleagues in Quito. It was possible for them to obtain advice by ham radio, but that meant divulging confidential information to eavesdroppers. When the men went to Quito to speak with the mission's Ecuadorian manager, he downplayed the lawsuit with such advice as, "Ignore the case. Don't worry about it. It's not important."

It was poor advice, but by the time Gordon and Art realized the importance of acting quickly, it was too late. Besides, they were dealing with a woman who had considerable influence in the area about Shell. The judge in the case told a friend of the hospital staff, "Of course, Victoria will win her suit: the missionaries are rich foreigners."

Art and Gordon spent valuable hours trying to settle the dispute out of court. The only lawyer in the area practiced in Ambato and came to Puyo only once a week. Besides keeping appointments with the lawyer, Gordon and Art visited the judge's office in Mera many times. When the case moved to a higher court, they traveled to Ambato.

Finally a settlement was reached in a third court where,

as predicted, Victoria won her case. The reason cited for the decision in Victoria's favor was that Gordon and Art had failed to appear in court when first summoned. The judge's ruling gave Victoria about one-fifth of the compensation for which she had originally asked.

Although the hospital lost the case in court, the decision was followed by an event that not only encouraged the missionaries, but also reassured them that God was in control. When mission officials arrived at the judge's office with the money awarded to Victoria, they met Miguel Guachamín, an Indian from the distant town of Loja. This man had sold some property to Victoria but had never received the payment she owed him. Now he had come to beg for that money. Certain that this encounter was no accident, and that Victoria would stubbornly refuse to pay him, the missionaries arranged with the judge to pay the Indian the amount owed him before the rest of the money was given to Victoria.

Despite the stresses of the court case, everyone at Shell rejoiced that God had not forgotten this poor Indian, who had now received his just due, or the missionaries themselves, whom He had sustained throughout the whole time of the trial.

18
Epidemic in Aucaland

"God is our refuge and strength, an ever present help in trouble."

– Psalm 46:1

In the jungle community of Tiweno, 30 minutes flying time from Shell, Rachel Saint and her Wycliffe partner, linguist Dr. Catherine Peeke, soon to be joined by a third co-worker, Rosi Jung, lived with a group of Aucas. With the help of Dayuma, their informant, the two missionaries had translated the Gospel of Mark into the Waorani language. While Catherine and Rachel did the translation, Dayuma taught her people the Word of God. As the work went forward, a remarkable change took place in the lives of the Auca Indians. Former killers became believers in Jesus Christ. Instead of mounting spearing raids and revenge killings, the Indians began to live at peace with one another.

The group of Indians at Tiweno, most of whom were Christians, was not large, yet these new believers were the nucleus of an indigenous church that was spiritually strong and possessed a missionary vision seldom found in older, more established churches. Farther downriver from Tiweno lived their relatives: people who had never heard the message of salvation. The Christian Aucas were anxious to make contact with their unreached kinsmen, and the knowledge that oil company workers were pushing farther into Auca territory had spurred them into launching missionary journeys. Their goal was to contact the groups that lived along the Curaray and Napo Rivers before any spearing raids against the outsiders could take place.

During 1968 and 1969, the Tiweno group journeyed inland, inviting hostile tribesmen to come and live with them so that they could teach them the Scriptures. By early August 1969 the Tiweno community more than doubled in number as down-river Aucas journeyed upriver to join them. Then, two weeks later, tragedy struck. Amoncawa, a husky young man, came down with an unknown disease: he was feverish, had difficulty breathing, drooled at the mouth and was partially paralyzed. He died the next day. Within a few days his mother died of the same dread disease, and a week later, a third adult died. Rachel, alarmed and puzzled, made radio contact with nurse Lois Pederson at Wycliffe's jungle base in Limoncocha and described the symptoms of the cases to her.

Lois could scarcely believe what she was hearing. "It sounds like polio to me," she said.

"Oh no! Surely it isn't polio!" Rachel was shocked.

"I think I'd better fly in there," Lois answered.

"You know the Aucas may spear to death any outsider who trespasses on their territory," warned Rachel, but Lois was determined to go.

The Aucas were normally a robust and healthy people; individuals died in spearing raids or from snakebites, but seldom from disease. But, unfortunately, since the Aucas had long lived in isolation from all foreigners, the tribe had no resistance to diseases common among outsiders. With the presence of so many unpredictable newcomers, the situation at Tiweno was tense, and Rachel feared that the polio deaths would provoke a reprisal. Hoping to prevent more cases, she sought medical advice by radio. In response, polio vaccine was flown from the United States and parachuted to her in the small jungle community.

Dr. Wally Swanson, then serving in Quito as director of HCJB's medical division, flew to Limoncocha and talked to

Rachel by radio from there; he agreed with Lois that the illness sounded like polio. In spite of Rachel's warnings, Wally felt that he and Lois should go into Tiweno to assess the seriousness of the epidemic. When they arrived, they were appalled to see so many sick people burning with fevers and struggling to breathe; in nearly every Auca home they found someone who was ill. Rachel, who had been up night and day caring for the sick, was near exhaustion; Lois decided to stay and help.

Wally returned to Limoncocha where he immediately set to work designing frames for rocker beds that would make it easier for the polio patients to breathe. When the beds were ready, he flew back to Tiweno where he and Lois strung hammocks from the crude frames and then hung them from the rafters of the roof of the Tiweno church. A person seated at the head of the frames could easily use a teeter-totter motion to keep the hammocks rocking--but someone always had to be there to keep the motion going. Between the extra nursing care that the epidemic demanded and the hours that they had to spend rocking the hammocks, Rachel and Lois became drowsy and often found themselves falling asleep.

On Wally's next trip into the jungle, he took a portable respirator, oxygen tanks and a suction machine. The medical equipment looked out of place in its primitive surroundings, but it was desperately needed. To save the life of one very sick young woman, Wally and Lois placed her on the electric respirator. During the night, however, Tiweno's generator broke down, the respirator ceased to function and the girl died. The next morning Wycliffe's pilot flew in a new generator.

Despite the best efforts of the tiny medical staff, the polio epidemic continued to spread. Nearly every other night a patient died. The deaths and fear of a hostile reaction from the unbelievers, who were known killers, drained Rachel and Lois emotionally. Since the two women were showing signs of exhaustion, Wally suggested that the sickest patients be transferred to Epp

Memorial Hospital in Shell.

Leaving Rachel and Lois working on a word list that would help the medical staff in Shell to communicate with their Auca patients, Wally flew back to Quito where he got permission from the Ecuadorian Health Ministry to use Epp Hospital as an isolation unit. Next he went to Baca Ortiz, the children's hospital in Quito, and asked to borrow the iron lung, a large steel cylinder that works on a bellows principle, pushing the patient's chest in and out to force respiration. The iron lung, the only one in Ecuador, belonged to Rimmer Memorial Hospital, but since the staff there rarely needed it, it had been loaned to Baca Ortiz. Knowing the desperate situation in Tiweno, the staff at Baca Ortiz gladly turned the iron lung over to Wally Swanson, and Gonzalo Ruales flew the machine to Shell. Next, Dr. Swanson went to HCJB's annual members' meeting which was then in session. He told the missionaries of the grave situation among the Aucas and asked for prayer and for volunteers to reinforce the Shell staff.

Dr. Peter Rae, a new missionary from Canada, immediately left the conference to travel overland to his post at Epp Memorial. Betty Harkins, the nurse in charge of in service education at Rimmer Hospital, accompanied Peter to Shell to help the staff set up an isolation unit.

Meanwhile, at the far end of Shell, Epp Hospital's maintenance crew lifted the cumbersome iron lung from Gonzalo's plane, loaded it onto a truck and then, trying to avoid the many potholes, drove along the bumpy road to the hospital. Once they had installed the huge machine in the largest room and had seen that it was functioning, the men breathed a sigh of relief, glad to have completed their nerve-wracking task.

On Thursday, October 2, 1969 the Wycliffe plane airlifted the first of 16 Aucas to Epp Memorial. By then, 14 of the tribe's adults had died of polio. When word came that more patients were being flown in, handyman Agustín carried his saw into the hospital, donned mask and gown, and then cut a hole into the adjoining room to enlarge the polio unit. Before he had swept up the sawdust, more patients had arrived.

At first the Aucas, who had never before left their home in the forest, hid under the bedsheets, but little by little their curiosity overcame their fears, and they began to take an interest in the activities around them. Since they refused the strange food of foreigners, the Wycliffe pilot brought smoked monkey meat, fish, plantain and yucca from the jungle for them to eat. Using an electric hot plate, nurses cooked the food in the isolation room and served it up, Amazonian style, on a banana leaf. Staff soon became adept at dodging orange seeds and other unwanted foodstuffs that the Aucas spit forcibly and unconcernedly across the room as if they were at home in the forest.

As news of the epidemic spread, medical personnel volunteered their services. Medical Assistance Program, Inc. in the United States arranged for Dr. Juan Correa, an anesthesiologist from the University of Oklahoma Medical Center, to fly to Ecuador in late October and spend ten days helping Dr. Rae. Wycliffe's new missionary, Rosi Jung, a German midwife, left her Spanish classes in Quito and went to Shell to help care for the Aucas. She stayed with the Wolframs, and each evening as she and Dorothy sipped tea, Rosi relaxed from the stress of caring for the Aucas, especially Nanca, who babbled in a language that he thought she understood.

During the epidemic, death came so suddenly! Though two nurses remained on the unit at all times, one of them could be calmly feeding a patient one minute and, the next, find that he had turned blue and was choking frantically on excess secretions. Two patients died, both from the downriver group. As they prepared to bury the first victim in the cemetery at Mera, the Epp staff arranged to take Dawa and Dyuwi, the two healthiest Aucas, along as eye witnesses. Afterwards, the two reported on the burial to their companions, assuring them that the foreigners had placed the bodies in a grave and covered them, and had not eaten them. Even so, the ways of the foreigners were strange to the Aucas who thought it better to be buried alive than to risk lack of burial that would send one's spirit wandering about the earth. Moreover, to die among strangers was chancy, for experience had proved to them that an enemy might leave a body for the buzzards and wild animals to eat.

The language barrier between nursing staff and patients put a great strain on nurses. Thankfully, a two-way radio hook-up in the polio unit enabled staff to talk with Rachel in Tiweno at any hour of the day or night. Rachel interpreted nursing procedures to the sick Aucas, thus relieving some of their anxiety.

Nanca and Enae were the two sickest survivors. Polio had left teenage Nanca a paraplegic, and nurse Betty Harkins was frustrated at her inability to communicate with him. Often she sat on the edge of his bed to show her concern for this boy, who would never again run through the forest or hunt monkeys. She was more than a little taken aback, however, when she learned that Auca marriages are arranged when two people sit in a hammock together!

Even during the epidemic there were things to laugh about. Betty was an excellent nurse. In the classroom she taught each procedure in detail. Thus, as she worked in the polio unit, she carefully draped Nanca with a sheet to avoid exposing him unduly during his bath. Then, remembering that she was not giving a demonstration to a class of nursing students, she asked her colleague, "Why am I being so particular? In the jungle this man wears nothing more than a G-string, if that!"

Enae, the other very sick patient, spent some three months in the iron lung. Nurses feared for her survival. On more than one occasion they did mouth-to-mouth resuscitation to revive her. When Enae was well enough to talk with the other Aucas, she recalled a dream in which she saw two trails, one broad and the other narrow. As she walked down the wide path toward a boiling lake, she saw people in it, screaming in agony. Frightened, she turned and ran back. Then she heard the two Aucas who had died call to her, "Enae, take the narrow path!" but when she tried, that trail was barred.

Enae told her story to Dawa, a Christian Auca, who explained that the wide path was the devil's trail. God had made another path, and it was through Jesus, whose way led to heaven.

"You were barred from entering that narrow path," her Christian companion told her, "because you have not yet let Jesus wash your heart clean." Enae understood this explanation and decided to become a Christian.

As the polio epidemic subsided, missionaries searched for a probable source of the disease. They traced its outbreak to a group of Quichuas who had not been vaccinated and had been in contact with outsiders. The Quichuas had stopped at the house of Amoncawa and had drunk *chicha* with him, using the same bowl. Another possible source may have been the Quichuas who lived across the river from the Aucas. Since the Aucas often raided Quichua homes, they could have carried off articles or food contaminated with the virus.

An estimated 95 percent of all the adults at Tiweno contracted some form of polio. A total of 16 died, and 12 others were left with some degree of paralysis. Auca culture made no provision for the handicapped. Only those who could hunt, fish and gather food had a right to live. Realizing this, staff organized a rehabilitation program. Dr. Wally Swanson contacted his brother Ellsworth, a physical therapist, who came to Ecuador and set up a physiotherapy unit in an Indian-style long house at Limoncocha. There he taught Lois Pederson and Rosi Jung the exercises that the recovered polio patients would need to practice. On Thursday, January 29, 1970 the doctors discharged Nanca and Enae, the last two polio patients to leave Epp Hospital.

For the next six months, 12 patients received therapy, and

all but Nanca recovered sufficiently to resume normal life in the jungle. Nanca was permanently hammock-ridden. Unable to hunt and fish with the other men, he made blowgun darts and traded them for food. Later on, he died, but Enae has survived. Despite numerous bouts of pneumonia, she still manages to care for herself and her grandchildren in Aucaland.

During the polio epidemic, Epp Hospital staff maintained a strict isolation regimen to prevent the spread of the disease. A separate staff of nurses cared for the polio victims, for instance, while a different group cared for the other inpatients. Nevertheless, in spite of these precautions, the number of hospitalized patients fell off. No surgeries were done, and people seemed reluctant to be admitted. In the months ahead, hospital admissions would remain low, for other factors, too, contributed to the decline.

19
Triumph in Tragedy

"Precious in the sight of the Lord is the death of his saints."

– Psalm 116:15

The search for oil in the *oriente* began with the unsuccessful efforts of the Shell Oil Company in the late 1930s. Yet despite the project's failure, Shell Oil did pave the way for future explorations. With the industrial countries' expanding demand for energy, oil became the world's most coveted raw material. In the 1960s, in Ecuador's northeastern jungle, Texaco discovered petroleum in sufficient quantities to make its recovery financially profitable.

Williams Company began the overwhelming task of building the oil pipeline. Out in the jungle, bulldozers and chain saws plowed relatively straight roads through virgin forest that only winding trails had penetrated before. Workmen crossed dense jungle forest and climbed across two mountain ranges divided by deep valleys to reach the shores of the Pacific Ocean. Though more than 40 of these men died in work-related accidents, Williams Company finished its task two months ahead of schedule. With the completion of the Trans-Andean pipeline from the oil fields to the Pacific coast, in 1972, Ecuador became a member of OPEC (Organization of Petroleum Exporting Countries). The discovery of "black gold" would greatly affect the life of the tribal people living east of the Andes.

With its excellent landing field, Shell became a strategic location for several major oil companies. Constant air traffic now broke the tranquillity of the area as commercial, mission-owned and oil-company planes roared in and out. Noisy helicopters lifted off the runway carrying prefabricated houses, trucks, metal pipes and various pieces of equipment to oil fields deep in the forest. *Río Amazonas*, Shell's airport, became the third busiest in Ecuador.

With all this increased activity, Shell became a boomtown. An influx of Ecuadorians seeking employment with foreign oil companies swelled the town's population. At the same time, down the road in the valley below the airstrip, a new company hired laborers to clear the land for planting fields of a tea-producing evergreen shrub. Epp Memorial Hospital played a vital part in caring for the health needs of workers from both the oil and tea companies.

Although, for the most part, the hospital treated patients with minor injuries, it would have cared for more serious cases if a resident missionary doctor had been available. The arrival of 55 new missionaries (regular, short-term and summer), however, and a difficult labor situation with HCJB's Ecuadorian employees had placed severe strain on the mission management. HCJB's senior members were overworked and loaded with extra responsibilities that stretched their energies to the limit. Thus, during 1970, the mission declared a moratorium on accepting new candidates and short-term workers that resulted in an acute shortage of missionary staff in most departments.

When Dr. Wally Swanson went on furlough in July, his departure left Epp and Rimmer Hospitals without a missionary physician. HCJB's two hospitals had to rely completely on Ecuadorian doctors, two of whom were hired for Shell. Dr. Hinojosa, the younger and a new graduate, was fulfilling his required year of rural medicine. Dr. Villacrés, the senior doctor, had had to leave his wife, a business woman, in Quito. Because she could not be with him in Shell, the doctor's schedule was arranged to allow him to be absent for several consecutive days each month, leaving his younger, less experienced colleague, whose professional and Christian ethics were questionable, without the supervision he needed. Friction developed between the two doctors. Each treated his own patients without communicating with or consulting the other; they even made separate rounds with the charge nurse. In this difficult situation, Epp's patient load decreased.

Yet God used these circumstances to lead a young doctor to apply for missionary service at Epp Memorial Hospital. Steve Nelson, a senior medical student, had spent the summer working with a physician in Cuzco, Peru. On his return trip, he stopped in Ecuador to visit Dave and Carol Osterhus of Mission Aviation Fellowship. As he shared with them his interest in missions, Dave and Carol said, "If you are interested in medical missions in South America, you don't need to go any farther than two blocks from here."

The next day Steve visited Epp Hospital. Unlike the average tourist, he talked with the staff and became acutely aware of the need for at least one missionary doctor at the hospital. At home again in Minneapolis, he contacted Wally Swanson. The two spent a delightful evening together discussing missionary medicine. During the next four years, while Steve was a medical student, intern, and a resident in family practice, thoughts of Shell and Epp Memorial Hospital receded into the background of his mind, but they never completely faded from his memory.

Meanwhile in Shell, health care improved when Ev Fuller and his wife, Liz, returned for six weeks in the summer. Year after year, the Fullers found themselves being drawn back to their roots in Shell. To help handle the many surgical cases, Ev brought Dr. Richard Cain, an anesthesiologist, with him.

The news that Dr. Fuller was attending patients spread quickly. From the surrounding area people poured into the outpatient clinic to see their beloved doctor. Scheduled surgeries increased, and with the presence of a surgeon, the number of patients soared. If the Fullers' visit seemed short, their contribution was of high quality. Those weeks were made up of long, busy, hectic but wonderful days filled with surgeries, emergencies--and fun.

Despite limited staff, there was no decrease in the number of visitors. Records show that during 1970, more than 350 foreigners came to see the work of Radio Station HCJB; most of these visitors made the long tiring trip by road, across the mountains to the jungle's edge. One group was a work team of 40 young people from Florida. In search of some cross-cultural encounters, these North American young people volunteered their services to the mission organizations in the area. Divided into four work parties, they painted the hangar at Mission Aviation Fellowship, the interior of Nate Saint Memorial School, the exterior of a house belonging to Gospel Missionary Union, and lightened the drab hospital corridor to a sparkling white. The experience gave them a wider vision for missions and personal spiritual growth.

That same summer an exchange group of teenagers from an Episcopalian church in Virginia arrived in Ecuador. A few days later, Bishop Onel Soto of Quito and Father Fisher of Virginia accompanied the young people on an outing to the *oriente*. Their two cars were traveling together along the precarious Shell-Baños road when an avalanche struck the second car broadside. The mass of rushing rocks and dirt sheared off the top of the jeep and threw the five passengers into a fast-flowing stream whose turbulent waters quickly swept them into the raging Pastaza River. Alerted by the noise of falling rocks, the passengers in the first car looked back to see their companions flung into the river. Three of the accident victims managed to climb out of the water. When friends tried to rescue the other two, they were able to pull the lifeless body of one 16-year-old from the river, but Becky DuPriest slipped from them and disappeared downstream. Father Fisher, driver of the second car, suffered from bruises and shock. He was admitted to Epp Hospital and kept under observation for several days. The other two survivors returned to Quito with the body of their friend, and staff at Rimmer Memorial Hospital prepared the body for transport and burial outside the country. Gazing sadly at the lifeless body of the teenager, the woman chaperoning the group said, "Becky and this girl were asleep in the back seat of our car; they never knew what happened."

Ten days later, a fisherman found Becky DuPriest's body lying on the sandy shores of the river three miles from the site of the accident. She was buried in Puyo. The cemetery there, the place of burial for both local Ecuadorians and foreigners, lies on a plateau overlooking the southern section of the city. As mourners and friends trudge along the dirt road and climb the steep steps leading to the cemetery, they look up toward a large open cross, cut out of concrete, that forms the cemetery's main entrance. Because of its elevation, the cross is visible for a considerable distance.

Just beyond this impressive entrance is the grave of another summer missionary. Jim Bucholtz, one of a group of 14 young visitors, traveled to Shell for a weekend outing one year later. The trip included a trek into the jungle station of Puyupungu at the juncture of the Pastaza and Puyo Rivers. Early on a Saturday morning, the young people squeezed into a small bus that to the tall North Americans seemed like a cracker box. Doubled up in the narrow seats and in high spirits over their adventure, they laughed and joked. At the end of the rutted road, a Quichua from Puyupungu came to take them on a three-hour hike that included crossing the river in canoes poled by Indians.

At Puyupungu, British missionary Mary Skinner welcomed the group. She was shocked, however, that no male missionary had accompanied them. Because the men of Puyupungu had gone to work for the oil company, leaving Mary to staff the station single-handed, she had asked for a male missionary to escort the group of young visitors. But that morning a breakdown of Epp Hospital's diesel generator had detained the two engineers who had planned to come with the group. Concerned, Mary warned the young people that the river was high, and that it would not be safe for them to go swimming in it.

The day was warm and sultry, and the young people, hot and sticky from their hike through the steamy rain forest, wanted to cool off. To them, the swift-flowing water looked inviting rather than dangerous. Fully dressed, they joined hands and stepped into cool refreshing water that came up only to their knees. Though Gail Allen, Jim Bucholtz and Becky Williams could not swim, they felt no fear as they joined hands with several others in the group and walked out into the river. They felt the strong current, but they were only curious when they saw some of their friends, farther downstream, climb onto the riverbank and hurry toward them. They did not realize that these friends were coming back to warn them of the danger, and so they ventured on. When the river suddenly deepened, one of the girls shouted, "This water is over our heads. We need to go back!"

Her cry came too late. Suddenly the young people were floundering in a strong undertow and swallowing water. Beneath its surface the rushing river had carved a 30-foot channel in the river bottom. Of the 14 young people, five were lifeguards, but when the swimmers tried to rescue the non-swimmers, they, too, nearly drowned. The young man who held Gail's hand was a strong swimmer. He tried to get her to tread water, but she panicked. Each time he went near her, she pulled him under. Frustrated, he left Gail and headed for Jim, but then returned to Gail. This time she was more cooperative, and he was able to tow her to the side of a perpendicular cliff where she clung to a rock until a canoe arrived. Within those few minutes, Jim Bucholtz drowned. Becky Williams was also caught in the strong current. She, too, would have drowned, if someone had not grabbed her dark brown ponytail and dragged her ashore.

When the young people realized that Jim was gone, they were stunned. They began at once to look for his body while Mary notified MAF by radio. The pilot flew above the area several times to search the river from the air, but he was unable to find Jim. What had begun as a happy outing had ended in tragedy. Jim had been so inspired by the unusual bus trip that he had said only that morning, "If we come back from this outing, the Lord must really have something special for our lives."

The Lord did have something special for Jim's life, though it was not what Jim had expected. Jim, who was only 20 years of age, was exceptionally talented in playing both piano and organ. The night before his death, in Shell, Jim had played several numbers on the small electric organ while the group sang; the organ music had never sounded better. At the same time, a prerecorded program on the theme "a new heaven and a new earth" was being aired on HCJB's TV Channel 4. On the program Jim played "Awake, Awake, for the Lord's Coming is at Hand" and "He the Pearly Gates Will Open." He did not guess that soon after the program aired, those pearly gates would open for him.

That evening the summer missionaries and the villagers held a service in the church at Puyupungu. Mary explained to the Indians that the visiting young people were sad at the loss of their friend, but that they were also experiencing peace and joy in their knowledge that he was with the Savior.

Two days later, Quichua divers found Jim's body at the bottom of the river. On July 6, 1971, Shell missionaries conducted a funeral service for Jim at Luz del Evangelio church, and memorial services were held in Quito and at Jim's home church in Ohio.

* * *

In 1971, following their second furlough, Dr. Wally Swanson and his family moved from Quito to Shell. Wally, a steady, methodical person, worked alone at Epp Hospital much of the time. Though he had no senior colleagues or specialists from whom to seek advice, he trusted the Great Physician, and his wife, Char, supported his medical work with prayer.

Her prayers were needed the day that the military stationed in Shell asked Dave Osterhus of MAF to make a three-hour emergency flight to bring a sick patient to Shell from Zumba. This small, isolated town lies 250 miles south of Shell, near the Peruvian border. Isolated from the rest of Ecuador by rugged mountains, Zumba is linked to the outside world by ham radio. The 950-foot runway, each end of which drops off steeply to lower ground, stretches across the top of a 4,700-foot high ridge. From that precarious landing strip Dave airlifted Segundo León to Epp Hospital.

Segundo, father of ten children, had had a ruptured appendix ten days earlier, but since the Leon family had no money for medical expenses, they simply waited, hoping that their father would get better. When, instead, the infection caused by his ruptured appendix spread through his abdomen, and it appeared that he would die, the townspeople offered to pay for an emergency flight.

When Segundo was admitted to Epp Hospital, his abdomen was swollen and tender. He was dehydrated, his pulse was weak, and his breathing was rapid and shallow. Though Dr. Swanson had little hope that Segundo would live, he started an intravenous drip with a strong antibiotic to combat the infection rampant in his patient's body. To save Segundo's life, surgery was necessary, and even at that the case seemed hopeless. Wally operated, excising the section of bowel that was dead, and then, concerned about Segundo's condition, stayed at the hospital all night to watch him, check his reaction to the medication and reread the available textbooks to make sure that he had not overlooked any means of helping him.

Since Epp Hospital had no blood bank, blood was drawn from donors and transfused immediately. All the pure-blooded Indians and many of the *mestizos* have O-type blood, and since Char Swanson's blood was also this universal type O, she often became a donor. At two o'clock in the morning, Wally went home, wakened Char and drew a pint of blood from her for the critically ill Segundo.

During coffee break the next morning, Char asked her weary husband, "How is Segundo? Do you think he will live?"

Tired and discouraged, Wally replied, "I don't think so." Then he added, "He probably wouldn't live even in the best university hospital in the United States."

Another day passed, and Wally was at the hospital almost constantly, but when he came home for a few minutes, Char asked, "Is Segundo still living?"

"Barely," Wally answered, though in their struggle to save the patient's life, he and the nurses were using all the equipment and medicines available.

Char knew that Segundo's grave condition was weighing heavily on Wally. That morning she began conversational prayer with her Savior. "Lord," she prayed, "if Segundo is not going to live, why did you bring him all the way from Zumba just to die here? If he isn't going to live, why did you let the poor townspeople there pay for his flight to Shell?"

In answer, the Lord reminded Char that her questions were like the plea of Mary, the sister of Lazarus, when she said to Jesus, "Lord, if you had been here, my brother would not have died" (John 11: 32b). Char believed that the answer to her questions was the same as Jesus' reply to the grieving Mary and Martha when He said to them, "Did I not tell you that if you believed, you would see the glory of God?" (John 11:40).

Later that day, Char went to Epp Hospital to see Segundo. He was more feeble than she had imagined.

Intravenous tubing ran to his right arm; a red rubber tube that drained his stomach of excess gases and secretions came out through his left nostril and gurgled its contents into a suction bottle near his bed. Taking Segundo's hand in hers, Char prayed aloud to God and then left the room.

The next day, eager to know how the Lord was answering her prayers, Char again asked Wally, "How is Segundo?"

"Better," Wally replied.

During his illness, Segundo was aware of the doctor's constant attention and the concern of the nursing staff. He even remembered the foreign lady who had entered his room and prayed for him. Never before had he met strangers who cared about him. By their actions rather than by their words, they were communicating Christ's love to a poor peasant from a small town on the Peruvian border.

Two months later Segundo, healed both physically and spiritually, was well enough to return to his family, who found him a changed man. Segundo's transformed life and his account of the excellent medical attention he had received from the staff at Epp Hospital opened the door for an outreach ministry. Members of Luz del Evangelio Church in Shell began missionary journeys to Zumba.

Meanwhile, Wally Swanson sought ways to improve health care at the jungle's edge. Often he gave a spinal anesthetic and then asked the most readily available person--a nurse's aide--to monitor the patient's vital signs while he operated. To be responsible for both the surgery and the anesthesia was stressful, however. Epp Memorial Hospital needed a competent anesthetist. Dr. Swanson now asked for one, a request that would bring more changes in HCJB's health care division.

20
Mending and Making Do

"May the favor of the Lord our God rest upon us; establish the work of our hands for us."

– Psalm 90:17

During HCJB's shortage of managers, mission leaders took on extra duties. Ralph Horn, who lived in Quito and managed one of the language departments, commuted between the capital city and Shell each week to attend to the most urgent administrative agenda at Epp Memorial Hospital. To help Ralph, the mission assigned David Roth, a new HCJB engineer to Shell, in the summer of 1971.

Dave, now in charge of Epp Hospital's maintenance, faced endless tasks, for before he could finish repairing one piece of machinery, he had to stop and fix something else. To begin with, the hospital cars needed an immediate overhaul, but before Dave had them running well, the two washing machines in Epp's laundry broke down. Then the diesel generator stopped functioning. Dave ordered new parts for the generator from the States, and in the meantime, Gospel Missionary Union's light plant supplied Epp Hospital with electricity.

One of the projects that had to be tackled at Epp that year was to set up a hospital kitchen. Since *Señora* Melba no longer raised pigs on hospital property, and since she had been serving meals to the inpatients for several years, the hospital hired her to run its kitchen. But staff soon discovered that animals fascinated Melba. It was not unusual to find birds, monkeys and chickens keeping her company in the kitchen. Repelled by this situation, Dave promptly organized an intensive clean-up campaign. First, he ousted Melba's pet monkey and parrot. Then he replaced the old wood stove with a new kerosene model that would not blacken the kitchen walls with soot. He had the workmen paint those walls a bright yellow, and he himself installed new lights. The hospital then hired two more women to help Melba, and Gail Allen, an HCJB nurse who had recently transferred to Shell from Rimmer Hospital in Quito, took overall charge of the kitchen.

Gail planned the menus, taught the three cooks nutrition and did her best to keep the kitchen clean. In one corner of the room was a tiny pantry that Gail thought would make an ideal spot for storing staples. Having scrubbed the pantry thoroughly and labeled the shelves, she was shocked a few days later to find four old hens flapping their wings and leaving their tracks on her once-clean shelves. When she asked the kitchen staff to explain the hens' presence, the women quickly replied, "But, Miss Gail, the maternity patients need their chicken soup!"

It was local custom for a woman to eat chicken for 40 days after giving birth to a baby. Since Shell had no butcher shop, vendors would sell live chickens to the cooks. Much to Gail's dismay, both boarding the fowl and slaughtering them left a messy kitchen.

Despite her efforts to keep the hospital kitchen well stocked, Gail soon realized that she could not buy meat in quantity in the local market. To overcome this difficulty, she ordered a side of beef from Macuma where Shuar Indians raised cattle. Assuming that the meat would arrive cut up and neatly packaged, she was flabbergasted, later that same afternoon, to find an enormous chunk of beef just where it had been delivered: right in the middle of the kitchen floor. She knew nothing about butchering, but to her intense relief, Morley Johnson, a GMU missionary who had some experience in butchering, returned from furlough that very day. With the aid of a cookbook diagram that showed various cuts of beef, he and Gail, along with several volunteers, began the tedious task of cutting, wrapping, labeling, and packing the meat into all the available freezers. In a tremendous spirit of cooperation, everyone worked late into the evening. It was well that Epp's staff had learned to be adaptable and to work together as a team!

In September, Sara Risser, nursing director of Rimmer Hospital in Quito, traveled to the *oriente* to evaluate Epp Hospital's nursing needs; her report would help the department head to make future staff assignments. Sara, a well-organized administrator, had written an agenda for her two days in Shell, but she soon found her plans thwarted. Epp's busy staff was rushing about taking care of emergency surgeries. Even Lois Price, Epp's nursing director, had no time to answer Sara's questions. Until Lois was free to discuss the nursing situation at Shell, Sara occupied herself with the mundane, but useful, chore of folding gauze into three-inch squares.

Sara spent most of two days at this task. While she folded, she noticed the many ways in which Epp's staff reduced unnecessary expenditures and controlled costs. In an area where most patients are poor, the nursing staff had learned to be resourceful. During slack times, they rolled cotton balls, cut up used IV tubing for drinking straws, hand-sharpened injection needles to be sterilized for re-use, and patched surgical gloves. Absolutely nothing was discarded or wasted. The doctors, too, did their part, managing with one tube of suture when most surgeons would have used two or more.

During her visit Sara also had an excellent opportunity to see Epp's need for a larger staff: a nurse was needed to organize the operating room and to standardize the surgical packs. Dr. Swanson made a request for someone to administer anesthetics, and Lois Price was asking to be relieved of her administrative duties.

The assignment of missionary personnel is never an easy decision for mission leaders, but when Sara returned to Quito, she recommended that Rimmer Memorial Hospital transfer Beth Huddleston and myself from its staff to fill the needs at Epp. During a furlough in Canada, Beth, a nurse with post-graduate training in operating room techniques and management, had learned from Dr. Jake Dick to give anesthetics. Under the direction of Jake and his colleagues, she had gained considerable experience in anesthesia. After her transfer to Shell, she would supervise Epp Hospital's operating room, be responsible for the preparation and sterilization of equipment, and administer the anesthetics during surgery; I would assume the duties of nursing director.

Beth and I arrived in Shell on December 29, 1971 and moved into the big house next to the hospital. The house was being remodeled into three apartments, but the carpenters, swamped with work, had not finished the remodeling. In the apartment assigned to us, the kitchen counter was a nauseating egg-yoke yellow, a color I disliked. Thinking that the hospital maintenance crew would install new Formica on the counter, Beth and I had bought this material in Quito and had it brought to Shell, only to discover that none of the hospital workmen had ever worked with Formica. I was dismayed, but Beth was undaunted. An only child who had spent as much time learning from her father as from her mother, she announced, "If the men down here don't know how to cover a counter with Formica, we'll do it ourselves!"

Borrowing a hand saw from the carpentry shop, Beth set to work. Since I knew nothing about carpentry, I followed her instructions in holding the material while she measured and sawed. Finally, after several hours, we had successfully laid new mottled-white Formica on a counter that was one-and-a-half inches wider at one end than at the other.

Beth was just as eager to update Epp Hospital's operating room, and she quickly saw the need for some structural changes that, when completed, would improve surgical working conditions. The sterilizing room, for instance, was a nightmare of disorganization. It contained a large gas autoclave, traction equipment, a barrel that overflowed with torn linen, and, in one dark corner, a sewing machine where the nurse's aides mended hospital linen. Although the room was small, and it was difficult to relieve the congestion, Beth was concerned that staff might accidently confuse unsterile supplies with sterile ones. To avoid any such possibility, she asked that the carpenters build a sliding door in the wall next to the sterile supply cupboard so that sterile packs could be passed directly from the autoclave to the cupboard.

Down the hall, in the center of the hospital, was another multi-purpose room. It served as a change-room for kitchen staff, storage space for oxygen tanks, and a bathing room for newborns, yet the linen cupboard, in constant use, was at the far end of the hall. The nurses affected by the remodeling campaign thought that it would be best to move the linen to the central room, ask members of the kitchen staff to change in their own work area, and put the babies in the same room as their mothers, where nurses could teach the mothers as they bathed the babies. Everyone enthusiastically supported this proposal.

Beth, however, had an additional concern: the delivery room. Despite her efforts to exclude visitors from an area that she considered private, curious people would stand on the outside steps leading to the upstairs apartment to watch the birth of a baby through the window of the delivery room. To put a stop to this practice, Beth added an item to the list of requested improvements: "Please obscure the delivery room windows by painting the glass white so that the women in labor can have some privacy!"

Unfortunately, Epp Hospital had a backlog of maintenance jobs. When the requests for renovations reached the desk of Fred Woodburn, who was filling in as administrator, he could only point out that the lack of systematic, long-term, programmed maintenance had overtaken the hospital's ability to fulfill its needs. The situation had reached the point where every job was a priority. Though Fred understood the needs inside the hospital, he knew, too, that the maintenance crew was busy constructing a new plant house to accommodate two diesel generators. Since the provision of electricity was an absolute necessity, the plant house had to be given priority.

At this point, Epp Hospital had only one operable diesel generator, and to conserve power, it had to be turned off at night. The night nurse did her desk work by the light of a kerosene lantern and used a flashlight when dispensing pills to patients. To provide sufficient light to work by in the emergency and delivery rooms and at the bedside of patients needing special attention, the ingenious Agustín rigged a car battery to power a light mounted on a mobile table that the night nurse wheeled from place to place. Once a second diesel generator was functioning, the kerosene lantern and the portable light could go into storage.

Eventually, the workmen finished the plant house, set up two diesel generators and moved the carpentry shop into the old generator building. The former shop became available for a storage space that doubled as a morgue. Yet that would change, too, for with no outside windows, the room was virtually free of bugs, providing an excellent place for the hospital's new telephone system. Robbed of a morgue, the resourceful nursing staff simply wheeled any dead body into the narrow hall alongside the radio room and hung a sheet across the entrance. If a death occurred during one of Dr. Fuller's visits, Beth would jokingly ask, "Why didn't you include a morgue when you built this hospital? Did you not expect anyone to die?"

The gray-haired doctor, by then older and wiser, would grin and say nothing. He knew that it was the nurse, not the doctor, who cleared the hospital corridor of curious onlookers while the distraught family wheeled the stretcher out into the corridor, where there was more space, and placed the body of their loved one in the wooden coffin the family provided.

It took months to renovate the hospital, and unforeseen interruptions constantly held up the work of the maintenance crew. One day, for instance, Silvia, a nurse's aide, found a snake lying behind some clothing in the linen cupboard. She paid no attention, thinking that it was a black electrical cord, but when the "cord" suddenly moved, she ran screaming from the room into the corridor. Staff summoned Agustín, but before he could kill the snake, it had slithered across the partition into the men's washroom. Agustín pursued it and killed it there, taking care, as always, not to destroy its head. Then he examined it and, turning to the apprehensive nurses, said, "Don't worry! This snake was not poisonous."

21
"Explosion, Exploration, Eruption!"

"Hear my prayer, O Lord; let my cry for help come to you . . . when I call, answer me quickly."

–Psalm 102:1, 2.

I was at home starting to make a pan of brownies when Luz María telephoned from the hospital. "Miss Beth says to come at once!" she said hurriedly. "There's an emergency Caesarian section."

The minor misadventures that Epp Memorial Hospital's staff encountered in improving the hospital's interior were pale indeed in comparison to the complicated medical problems regularly facing the Shell team. Many of Epp's patients were poor surgical risks, and both the anesthetist and the surgeon felt keenly the heavy responsibility of treating such difficult and complicated cases. Whether because of their weakened physical condition, the trauma they had suffered, or their delay in seeking medical attention, I knew that some patients were so gravely ill that no real hope for their recovery could be offered. On this occasion, nearly a year had passed since Beth had learned anesthesia, and during that time she had had no opportunity to practice the techniques. Now she would begin to anesthetize patients who were surgical risks from the moment they entered the operating room.

I turned off the heat under the pan of melting shortening and chocolate and rushed across the yard to the hospital. When I finished my surgical scrub and hurried into the operating room, Beth greeted me with the words, "Oh, Eleanor, please pray for me. *Señora* Ligia has almost no blood pressure, and now I can't find her pulse!"

Señora Ligia, whose preganacy was complicated by toxemia, was fast going into shock, for her blood pressure, 180 on admission, had plunged drastically; her unborn baby was dead, and she was bleeding profusely.

The day was warm and sultry. The air conditioner hummed in the corner window. Luz María turned off the noisy machine so that Beth could hear Ligia's heartbeat as she checked her vital signs, but in that instant a military plane roared down the runway for take-off, momentarily deafening us. While I arranged the sterile instruments on the table in preparation for major surgery, I prayed, like the psalmist, "Lord, help Beth, and answer quickly!"

From the instant that Beth injected the muscle relaxant to initiate a general anesthetic, she needed approximately thirty seconds to slip an endotracheal tube down the patient's throat. Since these are tense moments for the anesthetist, I prayed that Beth's first attempt would be successful, but Ligia's neck was fat and the opening narrow. When Beth rang the buzzer for assistance, Irenne came in and squeezed the black bag of the oxygen cylinder, forcing air into the patient's lungs while Beth tried once more. This time she succeeded.

Dr. Bryan Wigglesworth, a young English doctor with surgical training, performed the Caesarian section while Dr. Campana, his Ecuadorian counterpart, assisted, and I passed them the instruments. During the operation, our team struggled to save Ligia's life, but it was only by the grace of God that she came through the surgery. The Lord heard our hurried prayers and answered. That emergency was the first of a series of surgical cases. A week had passed before I could finish making those brownies!

In April 1972, Dr. John Doerfer, a surgeon and newly accepted candidate with HCJB, made a three-day visit to Shell. In anticipation of his visit, Dr. Wally Swanson had scheduled several patients for major surgery. The morning after John's arrival, he had just finished the first operation, and I was setting up the room for the next one when an explosion rocked the hospital. Down the hall, in an empty room, Venancio, Epp's cleaning boy, had plugged the electric polisher into a wall outlet to shine the freshly lacquered floor. As he connected the polisher, a spark ignited the volatile fumes of the fresh lacquer, causing a violent explosion and a flash fire. Administrator Fred Woodburn, a retired naval commander, tugged Venancio from the room, grabbed the fire hose hanging on the corridor wall, and put out the flames. In the same instant, Wally Swanson shouted to the staff, "Evacuate the patients!" Within minutes he and the nurses had everyone outside, even Pitur Puenchera, a patient whose fractured leg was in traction. Unable to wheel his bed through the narrow doorway, the staff loosed the traction apparatus and placed Pitur on a stretcher to carry him out of the building. The second operation had to be postponed for more than a hour while John dressed the burns on Venancio's face and hands, and Wally put Pitur back in traction.

The rest of the day went well, but that evening we were called back to the operating room because Rebeca Sandu, the first surgical patient of the morning, was hemorrhaging. John and Wally explored Rebeca's abdomen and found the bleeder. One hour later, as we walked out of the operating room, Char Swanson greeted us with the words, "Sangay is erupting!"

As she spoke, we saw the glare of red hot lava shooting skyward from the active volcano, silhouetted against the dark horizon. While we stood looking at this rare and awesome sight, Gary Quick, a medical student from the United States, commented, "What a day! Explosion, exploration, eruption!"

"Is it always like this?" asked Linda, a working visitor who was also from the United States. "I thought Shell was supposed to be dull!"

A few months earlier I would have agreed with her remark, but not at that moment or during the weeks that followed, for soon we would face more difficult surgical cases. One of these took place during the visit of another group of summer workers, exactly one year after the tragic drowning of Jim Bucholtz. This latter group arrived in Shell on a Friday afternoon and set out on their jungle trek to Puyupungu the next morning. It was a dismal, dreary day. The young people plodded through pouring rain, slipping and sliding in the mud. After their chaperones had described to them the tragedy of the previous year, not one young person dared ask permission to swim. But their weekend, which began in so dull a fashion, was about to change.

When the group returned to Shell late Saturday afternoon, the rain had stopped. The curtain of cloud and mist had lifted to reveal two snowcapped mountain peaks in all their splendor. From one of its three craters, Sangay, with its imposing pyramidal shape, was sending puffs of smoke skyward. Farther up the range, the setting sun bathed the ridge in a rosy glow, tinting El Altar's jagged summit a deep pink. The change in the weather lured the visiting young people outdoors with their cameras. Since they were scheduled to return to Quito the next day, the Swansons had invited all of them to their home for a social gathering that evening. I, too, had planned to attend.

At half-past six, however, the telephone rang. Beth, who was working the evening shift, was calling to tell me that I was needed to help with a Caesarian section. Irenne, who had worked the day shift, had already returned to the hospital to replace Beth so that she would be free to give the anesthetic. I hurried to the operating room. When I entered, Beth was at the head of the table preparing her medicines. Silvia, the circulating nurse's aide, was doing the pre-operative scrub on *Señora* Rosa, the pregnant woman.

Her enormous belly flopped back and forth like a ball on a volleyball court. Since the abdomen did not look normal, I asked Beth in English, "What's that bump--a hernia?"

"Yes," Beth replied, "That's a ventral hernia resulting from a previous Caesarian section; it has to be repaired, too."

I groaned inwardly but said nothing, for I knew that by the time the doctors did a Caesarian and a hernia repair, the party at the Swansons would be over.

"*Señora* Rosa's blood pressure is low," Beth said as she released the air from the cuff. "Pray for me."

As I set up my instrument table, I dismissed all thought of the evening's entertainment from my mind; silently I prayed.

Silvia and I were starting the sponge count when Dr. Swanson came into the room to pray with Rosa. Since patients facing surgery are anxious to be reassured that God is in control, a member of Epp's surgical team prays with them before an operation. Silvia and I paused while Wally prayed. Then he left the room for his surgical scrub. While he and Dr. Dave Munsom, a visiting pediatrician, scrubbed, they discussed the case.

Once the surgeon had made the incision and the surgery began, events moved rapidly. Soon we were calling out, "Suction, Silvia! Suction!"

Silvia ran across the room and flipped the switch on the noisy machine; Dave held the metal tip near the nick in the uterus, and at once the amniotic fluid gurgled through the tubing to the large bottle on the machine. Moments later, *Señora* Rosa delivered a baby girl. I handed Dave the infant aspirator and then glanced at the clock on the wall. It was 7:10 p.m., July 1, 1972.

Dave suctioned the mucous from the infant's throat; she uttered a weak cry, but her color remained a gray-blue. Usually Beth and the circulating nurse took care of the newborn infants, but when Dave started to hand the baby to Silvia, Beth said, "Someone else needs to care for this baby. The mother's condition is not good."

Sensing the gravity of the situation, Wally said to his assistant, "You had better step aside and take care of the newborn."

Dave stepped out of the sterile field, placed the baby in an incubator, and wheeled the equipment from the room. In a quieter atmosphere he and Irenne would give the baby oxygen. Meanwhile, in the operating room, the situation had become tense. Beth, at the head of the table and constantly in motion, struggled to stabilize her patient. *Señora* Rosa's blood pressure had dropped, and to increase the blood volume, Beth opened the valve on the intravenous apparatus. Then she started a second IV and connected a bottle of serum plasma. Next she checked Rosa's vital signs, pumping the bulb on the blood-pressure cuff. Lacking a monitor and equipment to give her automatic readings, she did manual checks. Silvia, at Beth's instruction, telephoned the lab, recorded each medication, prepared the next bottle of IV fluid and, since there was no monitoring equipment to guarantee an even flow, kept adjusting the fluid drips.

Zulay came from the lab and took a sample of Rosa's blood. Before long, Irenne opened the door a crack and handed Silvia a pint of fresh blood. We all exhaled as our anxiety subsided, for given the lack of a readily available blood bank, a pint of blood during an emergency was a precious item. Later, two more pints arrived. I knew that the blood donors must be the summer workers and thanked the Lord for His perfect timing.

Operations at Epp Hospital had a way of involving more people than just the designated surgical staff. Dave Findlay, filling in as hospital administrator, had informed the summer missionaries of the need for donated blood, and they responded at once. Dave spent the rest of his evening accompanying them, one at a time, to the hospital laboratory, where Zulay typed each person's blood until she found sufficient donors to fill the requests coming from the operating room. As, one by one, the group at the Swansons' dwindled, the party came to a quiet end.

When Wally finished suturing Rosa's uterus, he said to me, "The patient has signed the consent form for a tubal ligation." I handed him the necessary instruments, and he proceeded to do the minor surgery involved. Sterilization was becoming the most popular form of birth control, for many of the local women had been pregnant during most of their child-bearing years. With no opportunity to regain physical strength between pregnancies, their health suffered. Daily, women faced the problem of finding enough food and clothing to support their large families. The cost of tubal ligation was minimal if the operation were performed within a few hours of delivery, but Epp's doctors did not do the procedure without proper authorization. As Wally finished the ligation, he turned to Beth and asked, "How's the patient doing?"

"I can't get a blood-pressure reading," Beth answered, "but I have blood running now."

"Good," said Wally. "Run it as fast as it will go."

Beth found this more easily said than done, however. The intravenous needles kept coming out of Rosa's veins, first one and then the other. Dave returned to the operating room and, donning sterile gloves, did two surgical procedures to maintain continuous delivery of the IV fluids.

Señora Rosa had received medication to make her uterus contract, but instead, it remained flabby, and she continued to bleed. Wally massaged the uterus to stimulate contractions, but to no effect. Concerned, he said, "Please call Bryan. I think we may need to do a hysterectomy."

Bryan Wigglesworth was enjoying the singing at the Swansons' gathering for the summer missionaries. A trained vocal musician, he had been faced with choosing between a career in music and the medical profession, and though he had chosen the latter, he kept his interest in music. He came at once, and as he was entering the operating room, Wally asked him,

"Do you remember how to do a hysterectomy?" "I just looked it up," Bryan replied, setting the heavy surgical textbook on a near-by stool for easy reference.

When Bryan had scrubbed and I turned to gown and glove him, I caught a glimpse of Char Swanson standing in the corridor outside the operating room, eyes closed and hands uplifted in prayer. It was reassuring to know that others, too, were praying.

The next two hours were tense as Wally and Bryan clamped, cut and sutured, over and over again until they had removed the flabby uterus. Meanwhile, Beth struggled to keep *Señora* Rosa's heart rate and blood pressure stable. As the clock registered midnight, the surgical team finished the hysterectomy and began the hernia repair. At the same time, Rosa's condition stabilized, and we began to relax. The Lord had answered our prayers; our patient would live.

Down the hall, the nurses had changed shifts, with Gloria replacing Irenne who, by then, had worked six hours of overtime. Gloria, checking each patient, opened the door of the operating room and peered in, but on seeing that the surgery was still in progress, she closed it again.

When the doctors finished operating, Beth and I cleaned the room, leaving it ready for another surgery. It was 4 a.m. when we wheeled Rosa to the nurses' station where Gloria would care for her until morning. Then we headed home to sleep during the few hours that remained before dawn.

22
Bridge to the Community

"And [Jesus] sent them out to preach the kingdom of God and to heal the sick."

– Luke 9:2

Providing adequate staff to cover Epp Hospital's three shifts, its outpatient department, and an on-call schedule for emergency surgery was not easy. After working for two hours on nursing schedules, I decided to take a break. Besides, I wanted to visit *Señora* Rosa, the obstetrical patient whose critical condition had put us through some stressful hours two nights before. Although the nurses who were caring for Rosa reported that her postoperative recovery was satisfactory, I was interested in seeing her.

I found Rosa in her room, in bed, between clean white sheets. She was looking refreshed from her morning bath and holding her newborn baby cradled in her arms. Her black hair was parted off center and plaited in two long braids. A tiny mole dotted her forehead while a larger one blemished her chin. When I introduced myself, an appreciative look registered in her dark brown eyes; her round face lit up with a smile, revealing one gold-capped tooth and several gold inlays, typical of Ecuador's dentistry.

In chatting with Rosa I learned that she was a Christian; her husband had died at work, four months earlier, when a heavy metal pipe rolled from an oil company truck, striking him on the head. After her husband's death, Rosa and her two-year-old daughter, Judith, had moved in with Maruja, a married sister, the sister's husband and their nine children. Rosa's and Maruja's mother was also part of the household. Although Rosa appreciated the help of her extended family, she was the only believer among them, and she hungered for Christian fellowship. That morning she was rejoicing in a visit from the hospital chaplain.

Rosa spent 19 days in the hospital recovering from surgery. When Dr. Swanson wrote her discharge order, Rosa, who had no income, explained that she could not pay her hospital bill until a settlement could be reached with her late husband's employer. To make it easier for Rosa, Administrator Dave Findlay reduced part of her bill by using money that had been donated to Epp Hospital's charity fund. Dave also suggested that we take Rosa home--it would make it easier for him to locate her later, if necessary. Making an outing of the occasion, Dave took his family with him to deliver Rosa, and Beth and I went along.

The Findlays, with Beth and myself, piled into the hospital's jeep, while the hospital provided a taxi for Rosa and baby Carlota (named after Dr. Swanson's wife, Charlotte) and Maruja, who had come to visit Rosa. Following Maruja's directions, Dave drove into the country on a back road so narrow that the tall grass brushed the sides of our jeep. Suddenly Maruja called out, "Honk your horn several times!" The noise, she explained, would alert her family to our arrival.

Fields of sugar cane surrounded us; the tall grass concealed deep ditches that made roadside parking dangerous. Then Maruja pointed to a cleared area only large enough for a single vehicle. When Dave had parked the jeep, Maruja jumped out and reached into the tall grass to retrieve her long rubber boots. She put them on and disappeared along a path through the sugar cane, leaving us to await the arrival of a "jungle ambulance."

Soon two men appeared with a makeshift stretcher: a bedsheet knotted at opposite ends, hanging from a shoulder pole.

Placing Rosa inside the sheet, they carried her along a slimy path strewn with refuse. We followed, our sneakers squashing the sugar-cane pulp, which filled the air with a sweet odor of fermenting sugar. Along the way we encountered a single horse hauling stalks of sugar cane to the family press. Thirty minutes later we reached Maruja's house.

The exterior of the house, which stood on stilts, was made of unpainted wooden planks--black with damp. The interior consisted of a kitchen and two bedrooms. The family cooked their meals over an open fire built in a sandbox; nearby was the only item that reminded us of our own world: a shortwave radio. Deep in the heart of the sugar-cane country, this family had bought a radio so that they could listen to the programs broadcast by Radio Station HCJB. We sat in the only space available, on the edge of their beds. With the adults and children gathered round us, Dave read from his Bible, explaining the Scriptures so that even a child could understand. When it was time to leave, Maruja handed us a live chicken in gratitude for bringing her and Rosa home. We accepted her gift, knowing that to refuse would have offended her.

When we returned to Epp Hospital, Dr. Swanson was at the nurses' station, writing orders on the chart of a new admission. We told him about Rosa, where she lived, and the difficulty involved in traveling to and from the hospital.

"And to think I asked her sister to bring Rosa back in two days so that I could change her dressing!" Wally said. "Little do we realize the inconvenience that we sometimes cause our patients!"

Wally was in process of admitting four-year-old Gustavito, whose diagnosis was measles complicated with bronchial pneumonia. The three-room unit used to isolate patients with contagious diseases already contained a family of three adults and one child--all diagnosed with measles--but somehow the nurses stretched the space to admit Gustavito. During his admission process, the staff learned from Gustavito's mother that her husband had gone to bury another child who had died of measles, and that the week before a third child had died of the same disease.

Few children in the *oriente* were vaccinated against contagious diseases. Indeed, the lack of a vaccination program and preventive health education was a concern to the Ecuadorian government, as well as to the medical staff at Epp. Thus, in the early 1970s, Ecuador took steps to expand its medical services by providing health care to rural people. Every Ecuadorian doctor and nurse, on completing classroom training, would henceforth be required to spend a year practicing rural medicine. In Quito, Sara Risser, Rimmer Memorial Hospital's nursing director, and Carl Wilhelm, director of HCJB's medical division, kept abreast of developments in the national health system. Since the Shell hospital was considered an outpost, HCJB signed a contract with the national health department, and in September 1972, the government assigned to Epp Hospital Dr. Alonzo Cabezas and two Ecuadorian nurses, Edna Salas and Yolanda Reinoso.

During the new graduates' year of rural practice, the hospital's goal was to involve them in preventive health teaching. In 1972, Ecuador's primary cause of death was enteritis (inflammation of the intestines) and diseases relating to diarrhea caused by drinking contaminated water. Infant mortality was high: statistics showed that of every 1,000 children born in Ecuador, 90 died in infancy. If a child reached the age of one year, it merited a first-birthday celebration! Since so many patients were infested with intestinal parasites from drinking untreated river water, the challenge was to persuade women to boil drinking water for 20 minutes on a wood fire--no easy task without a clock, using wood that had to be carried from a distance. Educating the public would eliminate patients' returning to the hospital time and time again with the same medical problem. But the country people did not readily accept new ideas, and health teaching had to be repeated many times. The gap between the underprivileged and affordable health care was bridgeable, but it would require patience and consistent teaching.

In the hospital's outpatient department, the rural nurses began to offer prenatal classes, giving a prize to each expectant mother who attended six sessions. The nurses soon discovered, however, that most women made their first prenatal visit to a doctor in the eighth or ninth month of pregnancy, which meant that class attendance was low. Though the nurses dealt with a variety of topics in these classes, invariably the women drew them back to the subject of most interest to themselves: family planning. It was easy to understand the reason. One morning, for instance, a female patient pleaded with the doctor on his medical rounds, "What can I do, doctor? I have had 26 pregnancies. I have 10 living children, 10 dead, and I have had six miscarriages."

Although most women were keenly interested in family planning, their husbands did not always share their concern. One 30-year-old woman, exhausted after bearing 11 children and pregnant with her twelfth, came to see Dr. Swanson in the clinic, acompanied by her husband. After the doctor had taken her history, he discussed sterilization with the couple. The wife was willing to undergo the minor surgery required, but her husband refused to give his consent, for a large family would be the couple's only security during old age.

In this health-teaching ministry, the medical staff were crossing cultural barriers that are not easily bridged. One day, for example, in a Quichua village, the doctor examined a woman who was pregnant with her ninth child. When he discussed the possibility of sterilization, he learned that in the couple's thinking, every woman has 10 children inside her, and until the tenth one is born, no preventive measures should be taken to inhibit pregnancy.

Another time, when I arrived in an isolated village, Luis Ortega, who was carrying my equipment from the airstrip, pointed to his house and said, "*Señorita*, there's my house. I have had 20 children; 10 are living and 10 are dead".

Curious, I asked him, "*Señor* Luis, were all your children by the same woman?"

"Yes, *Señorita*," he replied, "by the same woman." Luis' wife, married at the age of 14, had been pregnant nearly all her adult life.

Anxious for the government to continue assigning rural personnel to Epp Memorial Hospital, Carl Wilhelm invited two doctors from Ecuador's national health ministry to visit Shell. As the doctors toured the mission hospital, they were amazed at its facilities for lab work and X-rays, the cleanliness of the rooms, and the systematic handling of clientele in the outpatient department. Moreover, they were dumbfounded to think that anyone would want to live and work at the edge of the rain forest to serve the area's medical needs.

Mission Aviation Fellowship flew Carl and the two visitors to Macuma to meet the health promoters. In 1969, in cooperation with Gospel Missionary Union, Dr. Wally Swanson had held a mini-course to train Indians chosen by their community to be primary health-care workers. Then, in 1971, Ella Rae Jones, a nurse, had instructed 21 Shuar Indians during a 10-day course. Since all other promoter programs in Ecuador had failed, the visiting doctors were surprised that this one was working, especially since the young men received no salary. After the visit, the officials from the national health department assured Carl that the Shell hospital would receive rural medical personnel for the coming year.

Having lived in Quito for 12 years, and being new to the *oriente*, I was astounded by the area's health needs. One day, as Joaquín Noroña, pastor of Luz del Evangelio church and chaplain at Epp Hospital, shared with me the church's outreach to isolated villages where the people rarely had a chance to hear a gospel message, I said to him, "Why don't we open rural clinics in partnership with the local church? It would give us the opportunity to vaccinate the children and to teach preventive medicine."

As a result of that conversation rural mobile medical clinics, often called "goodwill" caravans, came into being. Yet, in retrospect, Ev Fuller and Paul Roberts had been the first missionary doctors to be involved in rural clinics. Since then, Epp Hospital physicians, carrying their stethoscopes and a few medicines, had made infrequent trips into remote villages where they had held clinics, but the work had never involved a team effort. Epp's staff now began to organize a program that included health teaching, preventive and curative care, and a spiritual ministry.

This program required a tremendous amount of work to schedule the participating staff, arrange flights with MAF, prepare the medicines, pack a minimum of sterile supplies, buy food for the team's cook and, when the group had returned to Shell, submit a written report to both the local and national health offices. Communication with remote villages was highly restricted. If the village was not hooked up with MAF's radio network, the clinic's impending arrival was announced on HCJB's "Hymns of the Christian Life," a favorite early-morning program.

One of the first joint mobile clinics went to Sinaí, a community of about 60 families of homesteaders whom the government, three years before, had relocated from a mountain area to the southeastern jungle. Since Sinaí had no landing field, flying to the village was out of the question. An 18-minute flight in an MAF plane took the team to Chiguasa's grassy airstrip. From there, team members trekked for six hours along a jungle trail, panting as they climbed steep inclines and slid down muddy slopes. Risks were involved in traveling to and from such far-flung villages; trekking across rough trails through thick forest exhausted the team. Yet the clinics were a bridge to communities that had no other access to health care.

Through its ministry, the team showed Christ's love by

providing medical help and opened the door to evangelism. The villagers welcomed the team, for in their isolation, they felt abandoned. Though the government had provided the community with a schoolteacher, in the three years since he had been appointed, no one had gone in to supervise or encourage him. The village men made an occasional trip to the outside, but the women and children had not left the area since their arrival. Setting up clinic in the one-room schoolhouse, the team vaccinated the children and attended patients. In the evening, the whole community sat on backless benches while the chaplain preached.

On the team's second visit to Sinaí, MAF arranged for the Wycliffe plane to pick up the group at Chiguasa. When Wycliffe's single-engine plane took off with the clinic team on board, it failed to clear the treetops and crashed. No one was seriously hurt in the accident, but the aircraft's wings, propeller and wheels were damaged. The Ecuadorian Air Force's helicopter rescued the group and returned them to Shell.

23
Battling the Elements

"When you pass through the waters, I will be with you; and when you pass through the rivers, they will not sweep over you."

– Isaiah 43:2a

The September day dawned bright and cloudless. As the equatorial sun warmed the damp jungle air, the green slopes in the distance stood out in sharp contrast to the blue sky above. Such days in Shell are exhilarating, yet the weather can change suddenly and unpredictably, causing unanticipated difficulties. By mid-afternoon on this day, for instance, an ominous black cloud had descended on the mountains, and a torrential downpour was driving sheets of rain across the runway. Fierce winds rattled the hospital windows and whipped the surrounding trees into a frenzy, bending them earthward.

I stood with other members of Epp's staff at the nurses' station, listening to the rain pounding on the roof and looking out the window at the rainwater gushing from the hospital's overhanging eaves. Suddenly the violent winds lifted the aluminum roof off the carport and sent it rolling across the lawn, and in the next instant, a tree came crashing to the ground, just missing the high-tension wires.

Only moments before Irenne had said, "Someone should turn off the power in the light plant." But the force of the wind and the driving rain kept the workmen inside the maintenance shop. The new diesel generator had an automatic shut-off that would cut the power if an object touched its high-tension wires, but the mechanism had arrived from the manufacturer with the trigger installed backwards so that this safeguard did not work. The tropical storm was short-lived, however, and when it subsided and staff looked at the damage, they found that it was minimal. God had protected Epp Memorial Hospital and the area around it.

Out in the jungle heavy rains put many of the airstrips under water. On the next clinic trip to Sinaí the MAF pilot, unable to land on Chiguasa's waterlogged airstrip, flew the team to the solid runway at Macas. From there it was an hour's walk to the river where a cable car operated by Luis Malla carried the group halfway across the river to an island. Since a second cable car, connecting the island to the farther shore, was broken, Luis provided horses for travelers to ride.

The group mounted the horses, but as the animals plunged into the rising waters, the horse ridden by Lois Price, frightened by the rocks in the river, reared in panic and became unmanageable. Luis Malla, stripping to his underwear, waded into the river and, plunging through water up to his waist, led the frightened horse and its nervous rider to the other side. When the group returned from Sinaí four days later, the river had subsided enough for the team to cross it by stepping from stone to stone in the riverbed.

As Epp Memorial Hospital continued to provide rural mobile clinics, Margaret Gerbrandt, a nurse with Christian and Missionary Alliance, asked for a doctor to go with her on her trips to Quichua and Waorani villages lying along the Napo River. Epp Hospital provided the doctor, but high water and swift currents made river travel dangerous. Staff were grieved when they learned that an overloaded canoe carrying 29 Christians from the Pano church had capsized in a whirlpool on that river, drowning 11 people, including the wife of Santiago

Calapucha, the Quichua pastor who accompanied Margaret on these outings.

Through all the hazards, however, Margaret was optimistic. After one clinic, she wrote, "In spite of losing our way in the jungle and overturning our canoe in the Nushino River, we had a good week. To reach one remote village, we hiked across high ridges on a six-hour trek and then traveled three hours by canoe, but it was worth it. The doctor attended 43 patients, extracted several teeth, and at the evening service 12 young people said that they wished to receive Christ."

Closer to Shell, torrents of rain swept fallen trees and other debris into the swelling Pastaza River. During one storm, the bridge at Madre Tierra collapsed. The next day Beth and I drove to the river's edge to see the damage. Broken abutments lay upended in the rushing water. We watched fearfully as two courageous men crept cautiously along twisted railings and swaying cables. At any moment the rushing water could smash the broken abutments, snap the steel supports and send the men into the raging river. In spite of the risk, they made the crossing safely.

At another point, three brawny youths struggled to bring a cow across the river by canoe. To keep the frightened animal from drowning, one youth held its head above the water's surface, while his two companions fought to keep the canoe on course through the rapids. With each stroke of the paddle, the canoe, its three occupants and the young cow were swept farther downstream. When they finally reached the safety of the mainland, the strong current had carried them some distance downriver. Though the crossing was extremely dangerous, transporting the animal across the swollen river to the slaughterhouse probably made the difference between obtaining the money to repay a bank loan and losing the family farm.

The damaged bridge connected the mainland to an island, and on the far side, a barge carried passengers and cargo to the Sangay Tea Company and a farming area beyond. To people living in that area, the bridge was a lifeline, yet it was considered too costly to replace. Instead, the local people slung lines to support a hand-operated cable car that would carry them across the river. Nor was the cable car without its dangers: failure to let go of the cable the moment the platform gained momentum resulted in a number of crushed hands. Week after week, Dr. John Doerfer, surgeon at Epp Hospital, and his assistant repaired these injuries. Eventually, the cable car was motorized, but not until the doctors had reported to the authorities the large number of hand injuries it was causing.

Amazed by the number of hands needing surgical attention, John Doerfer did a study. From hospital records, he learned that during a 10-year period, a number of implements--machetes, sugar-cane presses, cable cars and firearms--had caused hand injuries, as had the illegal practice of fishing with dynamite. (The fishermen lit a stick of dynamite and threw it into the river to stun the fish, but if they held the stick a moment too long, the deadly charge exploded in their hand.) Dr. Doerfer's study showed that the machete, in constant use by tribal and country people, caused more hand wounds than did all the other implements combined.

Many other injuries were related to road conditions. Staff on duty at Epp Hospital were continually asked, "*¿Hay paso?*" ("Is the road open to traffic?") Epp staff were considered a reliable source of this information, for trucks rumbling by the hospital shook the building and rattled its windows. When the tremors were absent, staff knew that the main thoroughfare must be closed. Pounding rains often made the Shell-Baños road impassable, causing mud slides and serious bus accidents.

Official information on road conditions was available at Shell's military control post where soldiers checked all vehicles coming into the area. Since the road was the *oriente's* lifeline, its condition was everyone's concern. When the road was open, fresh eggs, fruits and vegetables brought by truck were available in the market. When the road was closed, people had to do without. The situation was difficult for Epp Hospital as well: medicines, oxygen for hospital use and bottled gas needed for cooking all came from Quito. Huge oil tankers from the coast brought fuel to the area, though when road conditions were poor, many truckers refused to take the risk of trying to reach the *oriente*, for they either carried no insurance or knew that their policy was invalid on roads where landslides were likely to occur.

In 1973, to replace the old dirt road between Pelileo and Baños, a road crew built a paved highway along the west bank of the Patate River. After dark one evening, a huge landslide struck a Baños bus, pushed it over the cliff, rolled it down the slope into the river and buried it beneath mounds of fallen rock and dirt. Except for a visiting North American Peace Corps worker, all the dead had lived in Baños.

The town of Baños, located at the foot of the volcano Tungurahua, nestles in a narrow pass of the Andes' eastern cordillera. Steep inclines and deep valleys conceal the mountain from the town's view. The people of Baños are Roman Catholic and deeply religious. Their cathedral is centrally located on the town's main plaza. Its interior walls portray vivid scenes that depict tragic happenings and incredible miracles said to have occurred in the surrounding canyons. Now, as the people mourned, a somber mood encompassed the area; men wore black ties and dark suits; women covered their heads with black veils and draped their shoulders with dark shawls. The construction of this new highway had brought more tragedies and no miracles.

The next year workmen tried to improve the Shell-Baños road, but frequent avalanches caused long lineups of cars and trucks waiting for the way to be cleared. When Epp Hospital staff received word that a bus and two cars had gone over the cliff, they prepared to receive casualties. Then word came that the vehicles were empty. Later that same week workmen recovered 10 bodies.

I worried every time a member of Epp's staff traveled out of the Shell area. One afternoon Lola, a nurses' aide, failed to report for duty. Planning to return to Shell in plenty of time to work her shift, Lola had gone to Ambato to give her daughters money to pay the fees charged to write their school exams. On her return, however, a huge mud slide had closed the road, making it necessary for her to spend the night in Baños. Next morning she and the other travelers plowed through mud up to their knees on the Baños-Shell road. At one point Lola felt herself sinking and cried for help; someone pulled her from the mud. When she and the bus passengers finally reached the other side of the mud slide, they faced another obstacle: the bridge was gone! Cautiously, they crept across a makeshift bridge of slippery logs to a waiting bus. Lola had adventures to describe when, finally, she reported for work.

With the road being repaired and closed much of the time, people traveled by air, yet poor weather conditions affected air travel, too. Thursday, May 2, 1974 was one of those on-and-off days: it rained, it cleared and then it rained again. It was my day off. Thinking that Sara Risser might fly down from Quito, I kept checking the weather. At 2 p.m. I watched as the DC-3 belonging to Aero Taxi Ecuatoriana Sociedad Anónima (ATESA) roared down the runway, took off and headed toward the mountains. By 4 o'clock that afternoon, since mist and fog shrouded the peaks, I gave up expecting Sara. At the same time, news was made public that Air Taxi, with three crew members and 22 passengers on board, was missing. The aircraft had never arrived in Ambato, and Shell's control tower had not heard from the pilot since he had passed over *Río Negro* (Black River). Given the weather conditions, air search for the missing plane was impossible.

Colonel Alejandro Duque, military governor of Pastaza, took command of the search operations. As he made preparations to handle a possible air disaster, he asked that Epp Hospital care for the casualties. As soon as the DC-3 was found, he would dispatch troops to patrol the area and to control distraught relatives and curious onlookers. The situation in Shell was tense. Most of the passengers on the flight, local people from Shell and Puyo, were known to everyone. We passed a somber night.

Next morning all ears were tuned to the activities on Shell's airstrip. At dawn an airplane took off: the search for the missing DC-3 had begun. Minutes later, the same plane returned; its pilot had spotted the wrecked DC-3 on the wooded slopes of Mount Añango near *Río Blanco*, only 12 minutes' flying time from Shell's airbase. A side wind had carried the aircraft 15 degrees off course and sent it crashing into the mountain, 300 feet from its peak. At once the Air Force's helicopter lifted off the runway and flew toward the pass. We canceled the morning's surgery and waited.

On the helicopter's return, rescue workers brought us two of the six survivors. Jorge Parra, a civil aviation employee, had suffered multiple fractures, and Darwin Montero, an 11-year-old from Puyo, had broken his left shoulder and right leg in the crash. Once the helicopter had left the two survivors with Epp's staff, it took off again and headed toward the pass, but this time clouds and mist obscured Añango's slope. The noisy machine went no farther than *Río Negro*, where it remained, unable to return to Shell because of the weather.

Epp's staff spent that day caring for Jorge and Darwin. During the long night before they were rescued, the two had listened to the groans of the dying, but by morning, all was quiet. Darwin kept repeating that his father had died in the crash. A woman passenger, realizing that she would not survive the night, had given Darwin her wristwatch to give to her immediate family. The cousin of nurse's aide Matilde León was on the same flight. Although Matilde was overjoyed to learn from Jorge and Darwin that her cousin was alive, bad weather prevented the Ecuadorian

Air Force from continuing rescue operations. Survivors left on the mountain slope would spend another day and night exposed to cold and rain.

Saturday dawned bright and clear. Before I was out of bed, I heard the planes; rescue operations had resumed. When I reached the hospital, the Ecuadorian staff, including off-duty personnel, was arriving. This disaster had touched the lives of everyone in the village, and the staff of Epp Memorial Hospital had responded admirably.

That morning the Rescue Squad brought us the three remaining survivors: Aída Duche, and Antonia Zambrano, passengers on the fatal flight, and Fausto Vilema, Air Taxi's cabin boy. Aída, Antonia, and Fausto were so cold that we wondered if they would ever be warm again. The two women described the horrible experience of knowing throughout the night that their fellow passengers, one of them Matilde's cousin, were dying. As Dr. Swanson took care of Fausto's injuries, the lad said, "Do you remember that I was in this hospital when Air Taxi crashed two years ago?"

"Surely," Wally replied, "God is trying to tell you something." But even though Fausto spent several days as a patient in Epp Hospital where the chaplain visited him, as far as we know, he never responded to the claims of Christ.

Since troops were no longer needed to patrol the hospital, Colonel Duque stationed them along Shell's runway to prevent the invasion of hundreds of persons who came to wait for the arrival of the crash victims. As we busied ourselves applying splints and starting IVs, we continued to listen to the depressing wail of the siren at the observation tower and the monotonous drone of the helicopters bringing in the bodies for identification. The residents of the Shell-Puyo area would mourn their loved ones for many months.

In the crash, Raúl Del Salto, pastor of the church in Puyo, died almost instantly. A graduate of Berean Bible Institute in Shell, Raúl became the first president of the Ecuadorian Evangelical Missionary Association (AMEE), burdened for the people of the Amazon region. Before boarding the plane, Raúl had been at Epp Hospital's pharmacy talking enthusiastically with Adriana about Puyo's forthcoming evangelistic campaign, of which he was the chairman. A true disciple of Christ, Raúl would be sadly missed by his family and the Christian community, for the *oriente* could boast few pastors of his caliber. It would now fall to others to take up the challenge and follow the path Raúl had chosen.

24
Perils of the Rain Forest

"The angel of the Lord encamps around those who fear him, and he delivers them."

– Psalm 34:7

When a runner arrived in Macuma with the news that Luis Jimbicti, leader of the Shuar Church Association, had been bitten on the leg by a bushmaster, Jim Hedlund of Gospel Missionary Union knew that he must act quickly. Poisonous snakes are more aggressive during their mating season, and since the bushmaster is venomous, prompt treatment of Luis' snakebite with antivenin serum was vital. Already three hours had been lost--the time it had taken the Indian runner to sprint through the rain forest--and delay could cost Luis his life.

Carol Osterhus was manning the radio at Mission Aviation Fellowship when Jim called her. Carol's husband, Dave, flying back from the north jungle, overheard the conversation and interrupted. "I should arrive in Shell in 10 minutes," he said. "Ask the hospital to send the antivenin serum to the hangar, and I'll take off again for Macuma as soon as I can pick it up."

When Carol telephoned the nurses' station, I was preparing patients' noon medications. Immediately I took two vials of the antivenin serum from the refrigerator, placed them in a bag, wrote a few words of instruction for dealing with snakebite and sent the package to MAF via Marcos, the cleaning boy.

Marcos arrived at MAF just as Dave taxied down the runway. Seizing the package that Marcos handed him, Dave at once took off again on the 25-minute flight to Macuma. When he was airborne, he opened the package and read the instructions I had written. Then he radioed Macuma and relayed them to Jim. One suggestion I had made was to apply ice to Luis' leg to reduce the swelling and slow his circulation. When Dave brought the aircraft to a standstill in Macuma, Jim, carrying a bag of ice cubes, jumped into the plane, saying, "We haven't time to deliver this on foot. I can direct you to the preacher's house."

Handing one vial of the serum to co-worker Joyce Stuck, Jim took off with Dave. Joyce hurried into the studio of *Radio Río Amazonas.* As the Shuar announcer watched her, she demonstrated how to administer the life-saving serum, and he repeated her instructions to his radio audience in his native tongue. Out in the jungle, Luis Jimbicti's family listened.

Meanwhile, in the plane, Jim pointed out Luis' home. Dave buzzed the house and then flew in a circle, skimming the roof as Jim dropped the ice cubes and serum at the Jimbictis' front door. Four hours had passed since Luis had been bitten, and seven people had been involved in getting the serum to him. Had it reached him in time, we wondered, or had our efforts been too late? Several days would pass before we would know the answer.

A week later rural nurse Cecilia Guevara and medical student Linda Londburg flew to Macuma to teach the Shuar health promoters. Joyce Stuck and her co-worker, Nettie Buhler, invited Cecilia and Linda to go with them to visit Luis. It was dark when, five hours later, the four women arrived at the Jimbicti home. Already some forty adults with their children and dogs had gathered to celebrate Luis' recovery by holding a "snake festival." Tired and covered with mud, the women washed in a nearby river. Then they entered the house to mingle with the other guests.

Luis' house, which was built of split bamboo and stood

on stilts, consisted of one large room and two smaller ones partitioned off for sleeping. Through the open doorway, Linda and Cecilia, who did not know the Shuar language, listened to the sounds of the group's animated conversation and watched the preparation of huge quantities of food.

Outside the house, the Shuar adults lit three bonfires. On one fire they singed a hairy monkey; on the other two they cooked plantain, corn and yucca in large pots covered with huge banana leaves for lids. When they had finished searing the monkey, they skinned it, tore it into chunks, and threw the chunks, one at a time, into a cooking pot.

As they did this they chanted the parts of the snake's body. Into the same pot they threw the heart of the chonta palm as a vegetable.

After eating a supper (served on a banana leaf) of monkey stew, corn, yucca, sugar cane and popcorn (supplied by Joyce), the four women went to see Luis. According to custom, he was in a separate room so that his children would not see him until he was well again. Only his wife was allowed to care for him. Luis was resting on a bamboo sleeping rack close to a fire that would warm his feet. Near the fire eight stones, laid out in the shape of a collar, signified that the snake was now tied.

Luis had received an injection of antivenin and the local witch doctor had treated him with a mixture made from the bark of a tree, hot pepper drinks and an enema of hot pepper. Joyce and Luis prayed together in the Shuar language, giving thanks to God that Luis was now out of danger.

The guests stayed the night, sleeping on the floor with the others. At three o'clock in the morning, to a background of radio music, the Indians beheaded a chicken and then prepared two special soups: chicken soup for Luis and monkey soup for the guests. At five o'clock, a messenger went into the forest to warn all the snakes not to bite any of the family or friends who were participating in this fiesta. When he returned, shafts of sunlight were piercing the forest's dense foliage, and the wild cacophony of bird calls had begun. While the women stirred the fires beneath the cooking pots, all the survivors of poisonous snakebites went into the sickroom. Laughing and making faces, they painted Luis' body with animal pictures, using a corncob soaked in dye made from a jungle plant. The snake festival ended with everyone feasting on more monkey meat, boiled yucca and plantain. As the four women prepared for their hike back to Macuma, the Indians took precautions to protect them from snakes by painting their necks with dye.

Meanwhile, in a small clearing, in another part of Ecuador's humid bug infested jungle, Erlinda Cuji lay injured, blood trickling down her face and hands. She had been helping her husband and father-in-law cut wood while the family's muzzleloader shotgun leaned against a stump, ready for use against any wild bird or animal. When the rifle began to slide, Erlinda reached out to prevent its falling, and the weapon discharged, wounding her hands and face.

The villagers were in a dilemma, for they had no health promoter and no communication system to ask for an emergency flight. But they did have Venancio, a Quichua pastor who had come from Arajuno to teach them the Word of God. As they gathered around the wounded Erlinda, Venancio reminded the villagers of a Scripture verse that they had been learning: "If you believe, you will receive whatever you ask for in prayer" (Matthew 21:22). Instructing the people to kneel, Venancio prayed that God would send help for Erlinda.

Meanwhile, Dave Osterhus sat in his plane on an airstrip only a six-minute flight from the village, studying his schedule for the following week. It was a Friday, and Monday's schedule was full. He realized that he should fly Venancio back to Arajuno that afternoon rather than waiting till Monday. He took off and flew over Erlinda's village, but thick clouds hid the tiny airstrip from his view.

Below, the praying Quichuas heard the plane passing over their village, but the sound of its motor faded into the distance. Dave had just radioed Carol in Shell to say that he was returning home when he tipped the aircraft's wing one last time. At that precise moment he saw a break in the clouds, and through the patch of clear air, he spotted the tiny airstrip hacked out of the jungle. Before the clouds closed in again, Dave slipped through them. His prompt arrival in their village in answer to their prayers made a deep impression on the Quichuas who were weak in their new faith and uncertain of the power of God.

Dave flew Erlinda to Shell where staff at Epp Memorial Hospital at once took her to the operating room. A surgeon repaired her wounds, and she recovered sufficiently to return home a few days later.

A few months after Erlinda's accident, I was a member of the first mobile medical team to enter her village. The request for the visit had come from Brethren missionary Lloyd Rogers. As the red-and-white Cessna made the sharp descent onto the grassy airstrip surrounded by jungle, Lloyd pointed to a shaggy thatch-roofed house on stilts. "There's my house," he said. "It cost me (the equivalent of $80 U.S.)" He went on to explain that since the villagers had had no money to cover the cost of Erlinda's surgery, he had paid her hospital bill; in return, to provide him with overnight accommodation whenever he was in their village, the Indians had built him a house of his own.

I stepped across logs, walked around tree stumps, and stumbled across uneven ground as I carried our equipment from the plane to Lloyd's house. Lisa Wingert, a short-term worker and our cook on this outing, nimbly climbed the notched pole that served as a stairway to Lloyd's only door. From below I handed her the cooking pots that we had brought with us, and soon she was busy at the camp stove, preparing our food as she sat with her legs dangling over the edge of the veranda. When two wooden tables appeared, I set up our equipment beneath the stilted house. Rural nurse Marta Gordón and I opened the village's first immunization program. We wanted to vaccinate the children who were six years of age and younger, but since none of the parents knew their children's ages, we had to guess.

The doctor on this week-long clinic was Steve Nelson. After his first visit to Shell in 1970, Steve had returned to the United States where he had completed his training in family practice. He arrived in Shell in May 1975, and now, in these primitive surroundings, he was adapting his North American medical training to a jungle setting. While the Quichua pastor served as interpreter, Marta helped Steve treat our patients, and I dispensed the medicines.

At sundown Steve, Marta and I joined the villagers in their new school. In the center of the room a single light bulb dangled from a pole. To provide electricity, our team had brought along a portable generator. The villagers were seeing an electric light for the first time, and in a few minutes they would see their first movie: a film that portrayed the life of Christ.

Early the next morning one of the villagers asked Dr. Nelson to attend a patient in labor. Taking along a minimum of equipment, I went with them to a house with a mud floor and a thatched roof, but no walls. Behind a plastic curtain, Amanda Santi, a timid young woman, squatted on the ground, ready to give birth to her first baby. Since Amanda did not know the doctor, she refused to let him examine her, and so Steve and I returned to the school where Marta was giving a class on the importance of cleanliness. It seemed ironic that she was teaching cleanliness when we had just been denied the opportunity to provide a sterile field for the birth of a baby! When Marta finished her lesson, I demonstrated the use of a toothbrush. Then we distributed a box of donated toothbrushes and paste. We had showed a second movie and were packing our equipment in preparation for flying to another village when Lloyd, known and accepted by the villagers, appeared to announce that he had helped at the birth of Amanda's baby!

Our team would hold several more clinics in this village where many would embrace the Christian faith. Rural mobile medical clinics were proving to be an effective tool in promoting preventive medicine and in presenting the gospel. Yet on some occasions the team realized that they were entering regions dominated by Satan.

On the second evening of a clinic in a village in the north jungle, Epp Hospital's mobile medical team was startled to hear several shots followed by the mournful wail of Indians. Grabbing an emergency kit and a flashlight, the group hurried in the direction of the wailing. To their horror they found Jaime Pisango, one of the believers, sprawled on the ground outside his home with seven bullets in his body. He was dead. Since the medical team could do nothing for him, they returned to the schoolhouse and went to bed.

Next morning the authorities flew into the village to investigate the murder. They learned that Jaime, a former witch doctor, had murdered someone before he became a Christian. The authorities thought that his assailant had been in the audience the night before when Jaime had made a public profession of faith in Christ. They considered his death a revenge killing.

Meanwhile, the Epp team took apart enough school benches to build a rough box that would serve as a coffin. When the authorities gave permission for Jaime's burial, Lloyd Rogers conducted a short funeral service, reminding the Indians that since Christ did not return evil for evil, neither should they. In accordance with Quichua custom, the family buried Jaime beneath his stilted house and abandoned the building. The weather-beaten frame house now became a resting place for the dead.

Opposition to rural mobile clinics was rare, yet on another occasion, the team encountered considerable animosity from a community of homesteaders living in the isolated village of La Séptima (The Seventh). The invitation to visit La Séptima came from the church leaders of a neighboring village where the people were eager to share the gospel. They provided horses for the trip, and one of the church members, Luis Ortega, went along.

The trail to La Séptima was rough and muddy; the horses were laden with boxes of medicines, the movie projector and a portable light plant. Along the way the group met a woman who, on realizing that the team members were evangelicals, became very angry. She cursed them vehemently, shouting that she hoped the horses would break their legs, and then went on her way. The group was stunned. Was it wise for them, they wondered, to go on after being cursed?

Jesús Montero, the hospital's chaplain, was with the group. He encouraged the team, saying, "God has not brought us this far for nothing. He will protect and bless us." Then he prayed, and the group continued its journey.

When the team came to a steep incline that led down to a river, the lead horse, which was carrying the movie projector and a box of medicines, plunged down the muddy slope, pitching violently from side to side, unable to find a secure foothold. The clinic team watched in horror! But the animal slid to a stop at the river's edge with the equipment still tied to its back.

Thoroughly unnerved by this event, but trusting the Lord to protect them, the group plodded on. Beyond the river, the way was blocked by a landslide. There was nothing for the team members to do but unload the horses and struggle on foot through the mud, carrying their equipment across the mass of dirt. When they came to a waterfall, they stood under it, letting the water splash over their hot bodies. From there the muddy path was laid with slippery logs, making walking even more difficult. The journey took a tiring three hours.

When the group had set up the clinic, many villagers took the opportunity to consult a doctor. While the clinic was in progress, the chaplain, as usual, made house-to-house visits, but in La Séptima housewives closed their doors in his face. He had never before experienced such cold hostility. That night, when he preached as planned, only a few children came inside the building. Outside, the adults crowded around the village guitarist who played loudly. When the guitarist finished, the crowd moved inside to watch a film on the life of Christ, but as they did so, men smoked and people talked. When the movie ended, *Señor* Montero again began to preach, but someone turned on a radio. When he asked that the music be turned off, the radio's owner raised the volume, making it impossible for the sermon to proceed. Jesús had never before encountered such disrespect.

The following morning the clinic team left La Séptima. They had gone only a short distance when they heard a shot and saw the bullet set leaves on nearby trees in motion. Unnerved, they ran, covering the distance on their return trip in half the time it had taken them to reach the village. When their plane landed in Shell, they were glad and thankful to be back at Epp Hospital.

But before Luis Ortega said goodbye to the group, he told them about an earlier visit that he had made to La Séptima. "Someone wakened me in the middle of the night to warn me that drunken men were gathering a supply of sticks and stones. They were planning to kill me. The man who warned me told me not to leave the village by the same way that I had come, and so, though I knew no other trail, I fled into the jungle. Behind me I could hear the shouts of angry men that grew fainter as I hurried on. By morning, I had reached home." Then he added, "I am certain that an unseen angel guided me through that snake-infested forest and brought me safely to my own doorstep."

25
Reaching Out

"Commit to the Lord whatever you do, and your plans will succeed."

– Proverbs 16:3

Rural mobile clinics from Epp Memorial Hospital continued to serve the isolated villages of Ecuador's *oriente* until the early 1980s. Although the clinics were effective in promoting good health in some areas of the rain forest, they could not meet the medical needs of every community. Missionaries with Gospel Missionary Union, found that the medical work was time consuming and often prevented them from performing their regular duties of opening churches and providing a radio ministry. Until 1977, Shuar Indians, who were being trained as health promoters among their own people, lacked consistent instruction and supervision.

Chuck Howard, Epp Hospital's administrator, met with the GMU missionaries several times to discuss the health care of the Indians. These missionaries wanted the young Shuar men who had been chosen as health promoters to take on the responsibility of looking after the physical needs of their own people, and HCJB's medical staff to supervise these promoters and undertake their ongoing education. If this plan were put into effect, the Shuar Indians would receive treatment for minor ailments in their own community and thus avoid unnecessary trips to the hospital.

Life would change for the person supervising the Shuar health promoters. To visit health workers living in far-flung villages of the Amazon rain forest, the supervisor would have to tramp along muddy trails, wearing rubber boots and carrying a backpack. Despite the physical demands and isolation that such an arrangement would entail, nurse Lois Price, who had been a staff member at Epp Hospital for 11 years, agreed to take on this duty. Lois related well to the simple country folk who lived in the rain forest's small communities. She had taught several health classes in Shell for the people of Río Chico (Little River) and San Jacinto, and while serving on medical mobile clinics, she had bravely endured the long rugged treks through the forest. Dr. Steve Nelson would help Lois handle this teaching ministry by dividing his time between Epp Hospital and the supervision of the health promoters. Steve's easy-going, affable personality and his ability to teach the young men in words that they understood made the Shuar feel comfortable with him.

When Lois left Shell for her new assignment at a jungle outpost, workmen were building a duplex on hospital property at the edge of the ravine. When completed, the duplex would provide more housing for nurses. Ed Duff, a retired building contractor and a widower, learned about the projected duplex from Chuck Howard when he visited Ecuador on an HCJB tour in 1976. To support the mission, Ed offered to oversee the construction of this new building.

Dwight Peterson, a United Airlines pilot, was also touring Ecuador at that time. Dwight had first become interested in Ecuador while he was a student at Wheaton College, a Christian university in Illinois, where he met Ed McCully and Jim Elliot, whose dream was to reach the lost tribes of South America. When Dwight learned of the projected duplex, he eagerly volunteered to help with the block work, providing that he could do so during his holidays. When the time came, Duff was able to give Dwight only two weeks' notice to join him in Shell. Though such short notice was unheard of, the airline granted Dwight's request for a vacation. His arrival in Ecuador coincided with the visit of Lee Johnson and his wife who were returning from a visit with their son in Brazil. During the next 10 days Dwight, Lee, Ed and the hospital workmen built the concrete-block walls of the duplex.

One evening the missionary staff sat in the Doerfers' living room drinking coffee and enjoying John's birthday cake. Chuck and Ed were discussing the need to install reinforcements in the new building to protect it in the event of an earthquake. Chuck, who was the son of HCJB missionary parents and had been brought up in Ecuador, said, "I think, Ed, that we need to build reinforcing rods into this building." But Ed was not convinced. Suddenly, amid the laughter and chatter, the house began to shake and the dishes began to rattle. The house, built by the side of the road, trembled with every bus and truck that rumbled by, but this was no truck; it was a severe earth tremor. When the shaking subsided, Ed leaned over and said to Chuck, "You win! We'll put in the reinforcing rods."

The men finished building the duplex in September 1977. The new building was dedicated on October 26 during the visit of HCJB's fall tour group. Beth Huddleston and I moved from the big house at the road to the quieter setting at the edge of the wooded ravine. Each morning we woke to the hum of countless insects and the twitter of jungle birds. Similar but more subdued sounds lulled us to sleep at night. I was delighted with glimpses of bright yellow-breasted birds flying across an empty space between the trees, and twice I watched as three marmosets ventured into the open to play and then, moments later, scampered back into the forest where jungle foliage gave them cover.

In 1976 when Gordon Wolfram and Chuck Howard staked the area for the proposed nurses' duplex, HCJB's board of trustees had no plans to do more building at Shell. Instead, mission leaders were working to expand HCJB's hydroelectric plant at Papallacta in order to provide more power for radio broadcasting. Yet even after the duplex was finished, Shell's staff found themselves coping with an expanding ministry in very cramped quarters. As a temporary measure--soon to become a permanent arrangement--the nurses squeezed a fifth bed into the four-bed wards. Carpenters partitioned off an alcove used by visitors so that it could become a nursery for premature babies. Newly admitted patients often had to remain in the hospital's hall until a bed became available.

The nurses' station was the hub of activity. Following surgery and deliveries, staff wheeled stretcher cases and newborns along the corridor to recover under the watchful eyes of the charge nurse--and in full view of curious visitors who inquired minutely about each patient's surgery. Meanwhile the charge nurse, who had to care for postoperative patients, answer visitors' questions, prepare bottles of intravenous fluid and check doctors' orders, also served as pharmacist and cashier after-hours. Opposite the desk and beside the time clock were the staff mail boxes. Mail from Quito was carried to Shell by staff traveling from the capital. When it arrived, missionaries gathered at the boxes, discussing the latest news from the homeland. In the midst of their hub-bub, the charge nurse gave her report to staff coming on duty, but the noise and distraction irritated the nurses.

Six days a week the outpatient clinic was bedlam. The expansion of the laboratory required moving the clinic nurse and her equipment into the hospital waiting room. Barricaded behind a counter, she checked her patients' vital signs in full view of the public; people in the crowded waiting room spilled into the narrow hallway. When the doctors stepped out of their offices, they had to run the gauntlet of a corridor of people waiting to see them and begging, "Please, doctor, see me next!"

Then Epp Memorial Hospital became a training center for Ecuadorian medical students during their year of rural practice. With the arrival of another missionary doctor, both he and the intern needed an examining room. Like players in a chess game, the hospital's administrative team constantly relocated staff and offices. The chaplain moved from the outpatient area to the room occupied by the hospital's administrator; the administrator, in turn, took over the nursing director's office, while her desk and files were relegated to an office on the second floor that used to be a storage space. A few months later, the chaplain moved again.

This time he went across the driveway to the old clinic, not a choice location for the person who had an important role in the hospital's spiritual ministry. Staff would see to it that his office was more accessible to the public, if and when a new hospital were opened.

In retrospect, it appeared that Epp Hospital had not always been busy. In the early 1970s the number of patients being treated dropped considerably. Several factors contributed to this decline. The departure of the foreign oil companies from the area affected the economy; the hospital lost clientele when the tea company hired a doctor for its employees; three Ecuadorian doctors in Puyo were giving free clinics, and Puyo's Social Security Hospital opened an outpatient department.

Epp Hospital's low patient count and the increase in employee salaries greatly concerned Hospital Administrator Norman Emery. Plagued by the ever-present problem of maintaining a jungle hospital in a healthy financial state, Emery thought that Epp's inpatient section should be closed, and that staff should concentrate on holding mobile medical clinics.

Not everyone agreed, for an important aspect of Epp's inpatient service was its spiritual ministry. One Sunday at Luz del Evangelio church, for instance, José Taco, a visitor from a distant village, stood up and testified that he had made a commitment to follow Christ while he was a patient in Epp Hospital. After he went home from the hospital, he and his family began to study the Bible together. "Now," he said joyfully, "six members of my immediate family are believers."

In Ecuadorian culture family members spend time with a sick relative. Knowing this, *Señor* Montero, Epp Hospital's chaplain, included the family in his ministry to the spiritual needs of each patient. One day he asked *Señor* Guacho, who was visiting his son Luis, "May I read a portion of Scripture to you?"

"Of course," was the reply. "We have plenty of time to listen to the Word of God."

In relating this story at an administrative meeting, Dr. Swanson said, "What better way to fill long hours of waiting at the bedside of a relative than listening to the Word of God!"

Although Epp's patient count remained low for several months, the situation began to change when the hospital acquired a surgeon and more full-time missionary staff. These people brought not only stability, but also patient confidence to the medical scene, and the number of hospitalized patients increased so that by 1977 it became necessary to hire someone to take on the secretarial duties at the nurses' station and someone to help the night nurse.

While Epp's administrative team dealt with the increased activity on the ward and in the outpatient department, down the hall Beth and Ena, her nurses' aide, shuffled surgical bundles in and out of small sterilizers. The big gas autoclave with a capacity for a large load had never been easy to light: either Beth or Ena had had to kneel on the floor beside it and, reaching underneath with a flaming torch, light the pilot light. Now it had stopped working completely. Beth adapted to the task of sterilizing large surgical packs without the big machine by dividing each pack into four smaller bundles that would fit into the portable autoclaves, but this approach quadrupled sterilizing time. To keep ahead of the demand for sterile supplies, Beth returned to the hospital after hours to run the sterilizers. When a new autoclave finally arrived 13 months later, it could sterilize in 45 minutes material that had taken a full day to sterilize in the small machines. The new machine, a big improvement from the obsolete model, functioned at the simple flip of an electric switch.

Fortunately, the new autoclave was working well when the doctors announced one day, "We are scheduling Sara Villalba for a total hip replacement. Dr. Uquillas, the orthopedic specialist, is coming from Quito to perform the operation."

Sara Villalba was the wife of Angel, the hospital carpenter, and the mother of Ena, a nurse's aide. The doctors had arranged to have her hip replacement done in Shell to save the family the expense of surgery in a city hospital. The concept was a good one, but the nurses at Epp were overwhelmed by the prospect. To prepare for surgery of this dimension, the nursing service arranged a combined educational and spiritual retreat for the nurse's aides. The aides were divided into two teams so that each team could be absent from its regular duties for one full day. Beth used textbook diagrams to explain the surgical procedure to the aides and then taught them the pre-and post-operative care required for a patient undergoing a total hip replacement.

Char Swanson, who was then living in Quito, arrived in Shell to help with the spiritual ministry to the aides during their retreat. She led the women in a Bible study and made herself available for counseling. As she dealt with the biblical meaning of forgiveness, the Lord brought tremendous blessing. Several members of the group were deeply moved.

One nurse's aide, whose husband had recently deserted her, cried and cried as Char pointed out the need to forgive, even if the offending person does not ask for pardon.

The second day's group was a little more subdued than the group that met the first day. Continuing to deal with the subject of forgiveness, Char asked them to kneel in a circle and to pray and ask forgiveness of one another. Unknown to the nurses, two people in this group were not on speaking terms, but as they knelt in prayer to ask forgiveness, God healed their wounds.

Later, they told Char that they were sure the nurses had arranged for them to be in the same group, but in reality no such arrangement had been made.

History was made at Epp Memorial when, on June 23, 1979, Dr. Uquillas performed a total hip replacement on Sara Villalba. The surgery went well, and *Señora* Sara made a full and uneventful recovery.

26
God's Open Door

"Go and make a survey of the land and write a description of it."

– Joshua 18:8b

With its Indian markets, snowcapped volcanoes, jungle lowlands and temperate climate, Ecuador attracts visitors year-round. When Dr. Nils Svensen, Per Danielsson and Gunnar Karlsson, members of HCJB World Radio's council in Sweden, visited Ecuador in January 1979, they came not as tourists, but rather to see the medical work with a view to funding ongoing and future projects. When the three Swedish visitors saw Shell's medical staff working in cramped quarters, they were convinced that the area needed an improved medical facility. They shared with Epp's staff their desire to solicit funds from the Swedish International Development Agency (SIDA) for the purpose of building a new hospital at the edge of the rain forest.

Epp's entire staff was ecstatic. Prospects for funding the project had come so unexpectedly! Just as the early villagers had felt the existing hospital to be a gift from heaven, so was the hope of a new facility. The Swedish delegation had enlarged the staff's vision and challenged their faith. With the help of an interpreter, Gunnar Karlsson said to me, "God wants to give you an open door that no man can close. He will help you resolve the details of a new hospital if you have faith."

That was the secret: "...if you have faith!" During the next two-and-a-half years, the faith of the Shell staff would indeed be tested. Until then, Shell's 1,400 inhabitants had lived behind storefronts and in tiny one- or two-room apartments above dull gray shops that were built literally wall against wall. To escape their dingy livingquarters, their occupants spilled into the street at daybreak. Rainwater from the roofs splattered into a ditch that ran between the wooden walkway and the dirt road. To the rear, tiny gardens sloped toward the winding Motolo River. Beyond the Motolo, several acres of land extended toward the north to the Pindo River. This wooded property, owned by a few long-time residents, served as pastureland for cattle. Since the military occupied all the territory south of Shell's main thoroughfare, the town could expand only to the north.

In the early 1960s, the local Catholic priest had encouraged the people of the area to form a cooperative. The municipality of Mera had divided the land into lots. For a monthly payment of as little as 20 sucres (then worth $1) local people who were interested became members of the cooperative and eligible for a draw that allotted them a plot of land for their use. Thus, those people who had the vision to invest in pastureland eventually owned property in a new subdivision that became known as Diez de Noviembre (The 10th of November), the founding date of Shell.

A year after the visit of the Swedes, Epp Memorial Hospital bought from Luis Tamayo and his wife, Elvia Santos, 16 acres of pastureland in Diez de Noviembre. The property was on the north side of the Motolo River, directly behind the hospital's existing premises and adjacent to the new housing cooperative, where five hospital employees already owned land. By 1980, the price of property in Diez de Noviembre had risen significantly. Since Epp Hospital lacked the funds to buy the land, HCJB's Shell missionaries pooled their resources and loaned the hospital the money. The staff planned to use this land for additional housing and to build a new hospital on the original site near the edge of the deep ravine.

To prepare for the development of the property, Administrator Chuck Howard arranged for Missionary TECH Team, an organization that helps missions with special projects, to survey the land. When Russ Chapman and Steve Richards, TECH's surveyors, finished their assignment, they met with Epp's staff. Impressed with the excellent location and quality of the new property, they challenged the staff to consider building the new hospital there instead of on the original site. The old site, they explained, would limit any future expansion, whereas the 16 acres in Diez de Noviembre could accommodate both the hospital and additional housing.

Taking the surveyors' advice, Epp's staff voted unanimously to build a 35-bed hospital on the new property. The townspeople were excited when they learned of the staff's decision. God was leading. To plan the design and layout of the new facility, Chuck Howard appointed an expansion advisory committee made up of Ecuadorian employees and expatriates Dr. Steve Nelson, Beth Huddleston and Dan Swanson, a certified nurse-anesthetist.

Meanwhile, hospital staff continued to care for the many sick patients arriving daily. One was Imelda, a 30-year-old Quichua woman eight months into her thirteenth pregnancy. When an evangelistic team of missionaries and Ecuadorians had visited her village in the Bobonaza valley, they had discovered Imelda in serious condition. She had been bitten by a poisonous snake six days earlier, and she was bleeding profusely. Since her body was covered with a rash caused by the snake's venom, but mistaken for measles, the Indian community would not allow her to pass through the village. To reach the airstrip while avoiding the village square, the evangelistic team carried Imelda around the outskirts of the town to the waiting plane. They were helped by a most reluctant husband who thought the effort a waste of time and money: his wife was dying; he could easily replace her.

When the team asked him how much he thought his wife was worth, Pedro replied, "No more than a thousand sucres [$25.]"

On Imelda's arrival, Epp Hospital staff realized that she needed blood, but her husband did not consider that she was worth a pint of his.

The lab technician could look for other donors. She did. Although the hospital staff gave Imelda six pints of blood, she continued to hemorrhage.

Dr. Jack Olinger, a physician experienced in internal medicine, and nurse Judy Cole were on duty that night. They felt helpless as they tried vainly to control Imelda's bleeding. Her unborn baby was dead, and she was fast going into shock. Unless she delivered the baby soon, she, too, would die. Because of the bleeding caused by the snake venom, surgery was out of the question. If the doctor and nurse could not help her physically, perhaps they could help spiritually, but neither spoke Imelda's language.

Judy called the home of Brethren missionaries Lloyd and Linda Rogers who usually had Quichua visitors staying in their home. In response to her call, 82-year-old Dr. Wilfred Tidmarsh --a geologist--arrived with five Quichuas. Having worked for years among the tribal people, he was making his last visit to Ecuador. For an hour, he and the Quichuas shared with Imelda their faith in Christ and their hope of life after death. Responding to their witness, Imelda believed.

Meanwhile Dr. Steve Nelson, arriving in Quito from furlough, would soon make a rapid adjustment from North American technology to jungle medicine. That evening he received a radio call from Jack Olinger who described to him Imelda's condition. Steve promised to consult at once with an obstetrician but found that he was pessimistic. Jack could induce labor if Epp Hospital had intravenous Pitocin on hand, but the obstetrician was sure that there was none in Ecuador; the Quito hospitals had only intramuscular Pitocin in stock.

When Jack signed off the radio contact with Steve, he walked to the little kitchen opposite the nurses' station, opened the door of the refrigerator and took out several ampules of Pitocin. All of them were mixed in oil for intramuscular use all of them, that is, except one. It was for intravenous use! Overjoyed at this discovery, Dr. Olinger and nurse Judy, at once started Imelda on a Pitocin drip.

Back in Quito, Steve Nelson spent the night alternately fretting and committing Imelda into God's hands. In Shell missionaries formed a prayer chain. About midnight Dr. Olinger noticed Imelda kneeling on her bed to deliver her baby, Quichua fashion. Realizing what was about to happen, he hurried home and returned with a rope, two nails and a hammer. From the hospital's wooden ceiling, he strung a rope for Imelda to hang onto as she bore down with each contraction. Once Imelda had delivered her dead baby, the bleeding stopped.

The next day Steve Nelson drove to Shell. During the five-hour trip he prepared himself for what he expected would be a sad story of a heartbreaking failed struggle to save a woman's life. Instead, to his amazement, he heard a story of joy and new birth, both physical and spiritual. Imelda had delivered her baby, her bleeding had stopped and she had accepted Jesus as her Savior--all in the space of 24 hours! Two days later, Imelda was walking and ready to return to her village: living proof of a miracle of God's intervention in her life. During the next several months, through her witness, this vibrant Christian saw her husband, three children, her two brothers and a sister become Christians.

* * *

In March 1981, two years after the visit of the Swedes, Missionary TECH Team wrote to Chuck Howard to ask if Epp Hospital was interested in having Ken Edgar, an experienced project director, oversee the building of a new hospital in Shell. Because of guerrilla activity in Guatemala, Ken was leaving a project there and would be available to supervise the construction. Chuck replied that the board would indeed be interested in this arrangement. He and the other staff at Shell had continued to be in touch with SIDA by ham radio, by letter and through representatives who came to Shell to discuss the new hospital's design.

A month later, the mission asked Chuck and his family to move to Quito to fill an urgent need in HCJB's accounting department. I was at home on furlough in Canada when Doug Peters, director of HCJB's health care division, telephoned me to ask if I would be Epp Hospital's administrator until Jim Estes returned from his furlough. I agreed, and then Doug, thinking that I might be overwhelmed by the assignment, said, "You won't have the hospital project to worry about. SIDA has turned down the hospital design that the Shell staff submitted."

I was stunned and disappointed for I knew that the expansion committee had worked hard on their design for a new hospital. Foreign mail was slow. I realized that I would not receive any further communication from Shell in the three weeks before I would return to Ecuador. All I could do was wait. But why, I wondered, had SIDA rejected Shell's design?

27
Problems, Provision, Praise

"Unless the Lord builds the house, its builders labor in vain."

– Psalm 127:1

Epp Hospital's expansion committee discussed and planned the new hospital with care, and then sent their design to the Swedish International Development Agency; but when Cecilia Phillipsson, SIDA's architect, saw the design, she rejected it. The expansion committee had planned a hospital that was one solid structure with a double corridor centered on a nurses' station. The committee had reasoned that a concentrated network of rooms would allow continuation of the existing arrangement of having two night nurses attend all Epp's patients. But Cecilia wanted a design that would provide more fresh air and light than a double corridor allowed. Furthermore, in recent years, SIDA had financed medical facilities in underdeveloped African countries where the hospitals consisted of complexes made up of several smaller buildings. SIDA representatives suggested a similar style for the new hospital at Shell.

In February 1981, a delegation from Sweden had traveled to Ecuador to meet with the Shell staff. Lengthy discussions had resulted in an impasse. The Shell staff knew that their design met Ecuador's standards for hospitals and was well adapted to the climate and culture of the *oriente*. Moreover, they were tired of waiting. Their mood fluctuated from elation to discouragement.

If SIDA did not approve their plan, they reasoned, it must be the Lord's will for them to look elsewhere for funding.

Then, six months later, after I had returned to Ecuador, HCJB's Swedish council arranged for Kyell Torbjornsson, a Swedish architectural consultant who was neutral to both parties, to meet with the Shell team. Project Director Ken Edgar spent the week in Shell, sharing a room with Kyell. The two men discussed means of incorporating into the plans for the new hospital all the features that the Shell staff considered necessary while meeting the concerns of the Swedes for providing a healthful facility. Ken realized that the next few days were critical.

When Kyell had met with Epp's staff and understood the area's medical needs, he sketched an entirely new plan. The hospital expansion committee immediately accepted Kyell's plan and, a little later, so did SIDA. Funds were now made available.

On January 4, 1982 Ken Edgar arrived in Shell with five workmen. The men were to remove certain trees and all underbrush from the building site. Because the project lacked an on-the-job supervisor, Herb Sevall, a retired road-building supervisor who was then visiting his son in Ecuador, agreed to oversee the work. Herb knew no Spanish, yet day after day, dressed in raincoat and rubber boots and holding an umbrella above his head, he managed the workmen.

Ken had said to me, "These workmen are to be paid each week." But the project had no accountant. Thus, in addition to my duties as Epp's administrator, I took on the accounting for the construction project. Six months later Wayne Wiley arrived in Ecuador to replace Herb as supervisor of the project, and--to my relief--that same month, Ken hired a bookkeeper for the construction work.

Heavy rains and spongy soil meant that the workmen had to build a solid road across the property before any construction materials could be brought to the building site. To do the rough grading on the main access road and to raise the grade of the lower end of the property, the municipality of Mera donated the services of a bulldozer for one week. The one thing that the workmen dreaded was rain which would hamper the bulldozing. It rained every day that week! In pouring rain the heavy machinery pushed the mud, first in one direction and then in the other. When the week ended, the bulldozer left, the rain stopped and the sun came out! Over the next several weeks, the workmen laid a foundation of river rock for the new roadbed.

Again, as had happened during the building of Epp Memorial Hospital, the Shell-Baños road caused problems. One week after the project began, this road was closed for major repairs. It remained under construction for an entire year, during which time it was open only from late Saturday night till early Tuesday morning. Travel to and from Shell was difficult. Huge oil tankers became stuck in the mud, and long lineups of traffic often delayed the delivery of building materials, planned for Mondays, until after the workmen had gone to bed. This meant that the workmen often had to be roused at night to unload materials so that the drivers of the delivery trucks could return to Baños before the road closed again on Tuesday mornings.

Ken Edgar, alternating his time between Quito and Shell, bought construction supplies, often spending hours looking for the best materials at the best price. He channeled working visitors to Shell, inviting electricians, plumbers, welders and carpenters from various countries to donate their much-appreciated expertise. In addition to the working visitors, full-time HCJB staff came to Shell to help in various ways. Each week I posted a new set of meal lists to provide for our guests, though road conditions still made their arrival times uncertain. One day Beth baked 60 muffins to serve an HCJB tour that never got beyond Baños. When the tour bus stopped there, someone told the driver that his bus would never reach Shell on the torn-up road. On hearing this discouraging report, the driver simply turned the bus around and took the group back to Quito.

Road construction brought with it many accidents. One evening two buses met on the narrow Shell-Baños road. As they scraped by each other, one went over the cliff, rolling down a 500-foot embankment and landing upright on the beach below. It took an hour for passengers on the bus that had remained on the road to reach the accident victims at the bottom of the cliff. Then darkness and a rough incline prevented rescuers from moving the injured until daylight so that 12 hours passed before the first patient reached Epp Hospital. Amazingly, only one life was lost--that of *Señora* León, mother of Matilde, who had been one of our nurses' aides. Only the week before, *Señora* León had waved at Beth and me as we drove through Mera.

Three days later word came that a Baños bus had been buried in a landslide. Two hospital workmen drove to the accident site, returning an hour later to report that although the bus was badly damaged, passengers had suffered only minor injuries. Another afternoon a huge mound of dirt and rock roared down the mountainside, catching Ramiro, a road construction employee, off guard, burying him up to his waist and leaving him embedded in the debris as if locked in a vise. Although his fellow workers acted quickly, by the time Ramiro was brought into Epp Hospital he was in severe shock from internal bleeding of a ruptured spleen; he also had a fractured pelvis. To save Ramiro's life, his friends donated several pints of blood, and the doctors performed emergency surgery.

While he was recovering, Ramiro and the hospital chaplain talked about life after death. Ramiro, who was not prepared to face life beyond the grave, argued that he knew Christians who did not live up to what they professed. Like many others, Ramiro was keeping his eyes fixed on fellow human beings and not on the Savior. Though he did not respond spiritually as the chaplain had hoped, the accident brought its blessing. The administrators of the expansion project paid Ramiro's hospital bill, and the road construction company donated crushed rock--an important item in block building--to the hospital.

Before workmen could finish the Shell-Baños road, floods at the coast destroyed more roads in Ecuador. When the construction company took its men and heavy machinery to the coast, the people of the Pastaza province protested; they did not want to be left dependent on a road that was only half finished. Agitators set up roadblocks, obstructing both entrances leading into Shell. Tensions ran high. Traffic in and out of the area ceased; schools closed; all construction work was ordered to stop, including the building of Hospital Vozandes-Shell.

Since buses were not running, Epp's patient load dropped. Still, the hospital needed to replenish supplies. Unaware of the severity of the situation, Wendy and I had driven to Quito to meet Beth, who was returning from her furlough in Canada, and to shop for hospital supplies. A few days later, when we were ready to return to Shell, all entrances to the Pastaza province were closed. Piedad, one of Epp's nurse's aides, who was also in Quito, had missed two days of work and was desperate to find transportation to Shell. Meanwhile in Shell, Hospital Administrator Jim Estes, newly back from furlough, arranged for Health Care Director Roger Reimer to drive us to the border between the provinces of Pastaza and Tungurahua, where he would meet us.

It was raining when our group reached the narrow bridge that spans the river dividing the two provinces. A dump truck, its tires deflated, surrounded with piles of stones, blocked the bridge. Most of the protesters were gathered inside a nearby building, though one or two stood guard at the bridge. Jim walked across the pile of stones to meet Roger.

"These guys are really tough," Jim said. "Some of them are communists, and they are armed. I'm going to ask again if I can bring three nurses and a nurse's aide into the province. If I come out of that building and walk toward my car, then you'll know that I've been refused, but if I walk toward the bridge, you'll know it's O.K."

Jim had tried at the military base in Shell to obtain a written safe-conduct certificate for four nurses to enter the province, but the commanding officer there would give him only verbal permission. Jim went into the building to talk with the protesters again; silently we prayed. The men were uncertain about allowing us to enter the area; they needed time, they told Jim, to think about his request. Then, just as Jim walked out of the building, the heavens opened, and a torrential downpour drove the guards inside. Jim hurried toward the bridge.

It was now or never! With sheets of rain driving against us, we carried boxes of medicines, food, three barrels of hospital supplies, tanks of oxygen and our suitcases across the slippery stones. Even Piedad, who was wearing high-heeled shoes, managed not to slip on the wet stones. When we had transferred all the supplies from Roger's car, he quickly turned around and sped away from the border; we did the same, heading in the opposite direction.

The road was not the only factor to cause difficulty in building our new hospital at the edge of the jungle. Six months after the project began, Epp received word from the administrators of the national power system that a high-tension line of 140,000 volts would pass across the housing area. A steel tower with a 12-foot-square concrete base would be placed against the rear wall of a family residence! The news was, to say the least, disturbing. Staff prayed that somehow the Lord would intervene. Meanwhile Ken Edgar negotiated with the power company. Its officials relented a little, consenting to move the tower several feet from the house in question. Although this concession was an improvement, none of the Epp staff were happy at the prospect of high-voltage wires passing across any part of the property. Then, unknown to the missionaries, Shell's municipal officials met with the national power company's administrators and persuaded them that the power lines should not cross any part of the new subdivision. Shell's intervention resulted in the decision of the power company to move the power lines 1,500 feet north, to the other side of the Pindo River, thus bypassing hospital land.

The pouring of the new hospital's 30,000-square-foot concrete roof was a major feat. Workmen hauled gravel and sand from the river. These materials were then sifted through screens, mixed with cement, hoisted to the roof by means of pulleys, and pushed in wheelbarrows to the points where it was poured. It took more than one major pour to complete the entire roof. The workmen, committed to getting this heavy job done, started early in the morning of one day and worked until two o'clock the following morning.

In Ecuadorian culture, the laying of a new building's cornerstone is almost as important a ceremony as its inauguration. The cornerstone for Hospital Vozandes-Shell was laid on March 21, 1982, just three months after the project began. Special guests attending the ceremony on that sunny Sunday afternoon included local dignitaries, military officers, and Christian Bausch, Sweden's ambassador to Ecuador. Quito missionaries, local church members and community friends swelled the audience to some two hundred fifty persons. The governor of the province and the local health officer commended Epp's staff for their vision of health care in the *oriente*. HCJB's Nordic Language Service provided special music, and Irenne Molina, Epp Hospital's director of nurses, spoke on behalf of the employees. Former hospital administrator Chuck Howard, who had done much in bringing the project to this stage, brought a devotional message based on the word, "Shalom," which may mean peace, health, abundance and victory.

"As we lay this first stone," Chuck said, "--a dead stone, so to speak--let us look to Jesus Christ, the Living Stone, and become living stones for Him." Then Sweden's ambassador and the governor of Pastaza province placed in the ground the "first stone" of the new Hospital Vozandes-Shell.

During the next three years, staff guided dozens of tour groups through the new hospital at various stages of its construction, explaining the purpose of each room. In August 1984, Karl-Eric Freed of HCJB's Swedish Council brought a group to Shell. Most of this group had followed the progress of the new facility from its early beginnings. When they saw the building nearing completion, their hearts were full of praise to God.

Pausing in the large, unfinished waiting room, Mr. Steig Wikströn, retired dean of the Swedish Alliance Mission Theological Seminary in Jonkoping, Sweden, said, "We feel that we are standing on holy ground." Then, opening his Bible to Psalm 118:23-24, he read: "'The Lord has done this, and it is marvelous in our eyes. This is the day the Lord has made; let us rejoice and be glad in it.' We are unable to attend the inauguration, and so today we want to dedicate this building. We want this hospital to be a place where Jesus will walk and people will feel His touch and be healed."

He went on to pray for all those who would work in the new hospital, that they might know God's calling in their lives. He prayed for the patients who would enter the hospital, that they might receive both physical and spiritual healing, and that they might sense the presence of Jesus as if He were present in His human body, moving from bed to bed. To close the service, the group sang in Swedish, "All the glory goes to Him. No human being can claim it."

Staff had planned to move into the new hospital on Epp's twenty-fifth anniversary, but May 10, 1983 and even May 10, 1984 came and went without marking the move. The next opening date was set for the fall of 1984, and although the two family residences, the four single units, the roads and sidewalks were all completed by that time, the interior of the hospital was not.

The days dragged by. Since few of the 60 workmen had had experience in construction, the men needed close supervision and careful guidance. Labor costs soared. Imported items were held up in customs for months at a stretch. The project was running out of funds. When the need was made known to members of Luz del Evangelio church, they took up a special offering that amounted to $1,000. This generous donation encouraged everyone and showed that the Christian community of Shell saw the need for the new hospital.

Finally, Jim Estes met with the project supervisor, and the two men set the date for moving: March 6, 1985; the official inauguration would take place three weeks later. Once this decision was made, everyone moved into high gear. Furloughs were postponed; no one wanted to miss this historic occasion.

At 1:30 p.m. on March 5, the day before the move, Epp's entire staff gathered at the nurses' station. In less than 24 hours they would move to the new hospital. They wanted to give God thanks for the privilege of serving Him within Epp's walls. Two long-time employees, Agustín and Luz María, in words that showed their deep emotion, described to their colleagues those early years, giving glory to God for His faithfulness. Brethren missionary David Butcher challenged the staff to seek new opportunities to serve the Lord in the new hospital. Maintenance Director Germán Rhon prayed, especially remembering patients who had found Christ as their Savior while they were hospitalized in that very building. Then everyone joined hands and sang the Spanish chorus, *Dios está aquí; qué precioso es* ("God is here; how precious He is!").

Unlike a staff transferring to a new medical facility in a developed country, Epp Hospital's staff was not stepping out of an old building and walking into a new, fully equipped, health-care center. Instead, the same furniture and equipment with which they had been working would be moved into the new building. For weeks the carpenter and painter were busy refinishing and painting hospital beds, bedside tables and chairs. The men worked hard, for when visitors saw the old items in place in the new hospital, they thought they were new.

The transfer from Epp Memorial Hospital's old building to the new Hospital Vozandes-Shell was to be a team effort, with all the mission organizations in the area and many of the local church people participating. To achieve a smooth transition from one building to the other, the doctors would do morning rounds as usual, nurses' aides would give patients their daily baths, and breakfasts would be served from the old kitchen; but all the patients would eat their noon meal in the new hospital. Apart from the cancellation of that Wednesday morning's outpatient clinic, medical services were to continue uninterrupted. Every employee was scheduled to work on March 6, the day of the move, and each knew his or her assignment. Missionary wives volunteered to prepare food for the mid-morning coffee break, and the schoolteachers planned a picnic to keep the children, who usually went home for lunch, occupied during the noon hour.

In the hours before the move, the administrative team passed through some anxious moments: the larger diesel generator had been installed at the new site, but it had not yet been hooked up to provide automatic back-up electricity in case the local power supply should fail. The new telephone system was not working. To move to a new building that lacked these necessities could jeopardize a patient's life. Although Administrator Jim Estes had asked for help from HCJB's engineers in Quito, he had received no reply. Had administration made an error in transmitting the target date? By now the local radio station was announcing the hospital's relocation, and the date of the official inauguration was public knowledge.

Believing that God was in control, Epp's staff decided to proceed as planned. Then, just before noon on March 5, three of HCJB's engineers walked in the back door of Epp Hospital, announcing their arrival. Within a few hours, they would hook up the electric circuit and the telephone system in the new Hospital Vozandes-Shell.

28
The New Link

"The glory of this present house will be greater than the glory of the former house."

– Haggai 2:9a

The morning of March 6, 1985 the people of Shell awoke to the patter of raindrops and a feel of dampness in the early morning air. Because of the hectic day facing them, the missionaries prayed for the rain to stop, and by eight o'clock the weather had cleared and the big move began. Beds, bedside tables, wheelchairs, stretchers and incubators were rolled down the cement ramp at the rear of Epp Hospital for loading into trucks and private cars. While Zulay and Adriana boxed equipment and medicines in the laboratory and pharmacy, Wendy Freeman labeled the cartons for easy identification and Jim Estes loaded them at the hospital's front entrance. Cartons and cartons of patients' charts, neatly arranged and numbered by the night watchman who had dismantled the chart room during the preceding night, lined the outpatient department's corridor.

To prepare for this day, volunteers from the missionary community and Ecuadorian employees had spent the last four evenings in the new hospital, washing walls, cleaning windows, scrubbing counters and sinks, waxing and polishing floors, hanging curtains and drilling holes for towel racks, fire extinguishers, soap dispensers and pictures.

But now, on the morning of the move, the new hospital

was unexpectedly alive with workmen. Lights flashed and buzzers sounded as the electrician made his final check. In the outpatient department the carpenter's hammer pounded nails into a new chart rack, and construction workers moved in and out of rooms.

Beth Huddleston, who had been assigned the task of directing moving-day traffic in the new building, was surprised to find the place overrun with workmen. She went directly to the operating room, where to her dismay, she found scaffolding still in position as it had been left to support the crew that had installed the ceiling light. Two painters, standing on the scaffolding, were touching up the paint around the light.

"When will you men be finished?" Beth asked the painters.

"*Ya mismo*," they replied. The words mean "Any time now," but in Ecuador that can mean any period from a few minutes to several hours.

"When is "*Ya mismo*"? Beth pressed them. "In an hour?"

"*Quizás?*" ("Perhaps"), they responded.

Beth went to the second operating room and found a painter occupied at the same task. Next she hurried to emergency, and there was another painter touching up the ceiling. By this time equipment from Epp Hospital had begun to arrive at all three entrances of Hospital Vozandes-Shell. Rushing from entrance to entrance, vainly trying to be in three places at once, Beth paused long enough to call me on the telephone. She sounded exasperated.

"Eleanor, don't send the operating room equipment until I clear the workmen from the surgical unit!" Uppermost in her mind was the knowledge that at any moment either the old or new operating room might be needed urgently for emergency surgery. Taking her advice, I halted the dismantling of the old operating room until further notice.

My next call came from Jim, who was speaking from the

far end of the old building. "Eleanor, I'll be out of the picture for awhile. We have a flood at our house. A water pipe has broken."

I assured Jim that the others could continue to pack and load until he returned.

MAF called next. That morning the pilots, all of them expert mechanics, had discovered that every plane but one was grounded with mechanical problems. Since the pilots were unable to help directly with the move, they had loaned the hospital their car and driver, but now they needed the car to bring a patient from the hangar to the emergency room. Nestor Kajekay, a Shuar Indian, had fallen against a rock during a drunken brawl and fractured both his forearms. When he arrived at Epp Hospital, the staff had almost finished dismantling the emergency room. To take pictures of his broken arms, Dr. Nelson used the portable X-ray machine and then went to the new hospital to develop the film.

At 10:00 a.m., still on schedule, I sent Beth more staff members. By 11:00 a.m., the nurses had finished their patients' routine morning care and were ready to transfer them. Using the hospital van that easily accommodated stretcher cases, Dr. Steve Nelson and Dr. Bruce Mertz moved the inpatients. Since Azucena Carvajal, one of the patients, had had major surgery the day before, she was given an injection to relieve pain about twenty minutes before she was lifted from her bed to make the uncomfortable journey, and Steve drove slowly and carefully through the town of Shell. Just ahead of the van, local workmen, in response to our request made earlier that week--were filling in the many potholes to reduce the bumps in the road.

When Steve and Bruce finished transferring patients, the only patient remaining in the old hospital was Nestor Kajekay with his broken arms. The emergency room nurse had gone to the new hospital to begin attending patients there. She did leave enough cast material at Epp Hospital for Dr. Nelson to immobilize Nestor's forearms. When Steve finished applying the casts, he and Nestor left. Of all the staff, only Luz María and Lidia Grefa, the cleaning woman, remained in the building with me.

The old hospital, now empty, seemed barren and desolate. I walked down the silent, gloomy corridor. The rooms were full of memories. "If only these walls could speak, what stories they would tell!" I said to Kathy Jo Estes, who had come to help clean.

I paused outside the empty pediatric ward. It had seldom been vacant; indeed, at night it had usually been overcrowded, with a mother sleeping on a thin mattress beside each child's cot. Rainbowbright colors still covered one wall of the unit. Cleverly sketched monkeys, macaws and parrots brought the jungle to life on the other walls. The pictures had been painted by a group of summer missionaries to portray the story of Noah and the ark. Before these young people had put down their paintbrushes that summer, Noah appeared on the door to pediatrics in the guise of a mountain Quichua wearing a typical dark poncho. Hospital visitors often stopped to admire the scenes.

One child whom I especially remembered was four-year-old Efrén Mesías, who was suffering from a parasitic infection and a low blood count. When Dr. Alonzo Cabezas saw Efrén's lab report, he asked the child's mother to donate a pint of blood for her son. When she had done so, Dr. Cabezas admitted Efrén to the pediatric ward, and his mother, carrying a younger child in her arms, left the hospital never to return. Efrén improved rapidly and, a week later, was ready to go home, but no one came to visit or collect him. After the second and third week had passed, staff knew that this little boy had been abandoned. During the next five months, Administrator Norman Emery tried to locate Efrén's mother, but when all his efforts failed, he did the necessary paperwork to have Efrén admitted to a Christian orphanage. That year Efrén spent Christmas at the hospital, and when a caring family gave him new trousers and a shirt, he happily showed them off to the staff.

I left pediatrics and walked across the hall to a room that reminded me of Marcelino Wambutzara, an emaciated Shuar teenager with a skeleton-like figure. Marcelino had the worst case of tuberculosis that our doctors had ever seen. His condition was complicated by multiple large cavities in a persistently collapsed lung. Besides being physically ill, Marcelino suffered from loneliness and depression. Highly infectious, he was isolated from other patients, and none of his family ever came out of the jungle to visit him. Yet God had met him in his need through the ministry of the hospital chaplain. We hoped, too, that his health would improve. Then, quite unexpectedly, our hopes for Marcelino were shattered when he developed aplastic anemia (fatal damage to the bone marrow), a side effect of his anti-TB drugs. This sudden change was difficult for everyone. I longed to help Marcelino in his dying hours, but there was little I could do. When he died, we grieved over the loss of a friendship that began months earlier. Yet we knew that in heaven the angels rejoiced as the Savior welcomed Marcelino to his eternal home.

I walked down the hall to the men's ward, remembering Rafael Mashient, the pale young Shuar health promoter who, ill with leukemia, had occupied the bed by the window. Anxious to learn more about his weakened condition, Rafael asked his nurse, "Can you tell me about leukemia?" By his third admission to Epp Hospital, Rafael knew that he did not have long to live. When he was leaving, he pointed upwards as he said to Dr. Nelson, "I'll see you up there."

Moving from room to room, I helped Luz María and Lidia sweep the floors, by then littered with bits of adhesive tape, empty cardboard boxes and discarded memos from the bulletin board. As we passed through patients' rooms, we removed old bed numbers and rusted thumbtacks from the wooden walls. Scattered over the floors were piles of fine sawdust left by the destructive termites that infested the frame building.

When we had finished cleaning, Luz and I sat down at the empty nurses' station. We both felt reluctant to leave the building where we had spent the best years of our lives. Associations and memories haunted the place. Since most of the Ecuadorian staff then lived either in town or in the subdivision, only Luz María and one other person would occupy the second floor of the old hospital. With the closure of this health facility, the closely knit family unit of former days would be gone.

As we reminisced, the nurses' aides arrived, one by one, from the new hospital where they had finished their shift. "We wanted to see what the old hospital looks like without patients," they announced. Then, bubbling with excitement, they told us what was happening across the river at the new Hospital Vozandes-Shell. Julia Morocho, who had gone into labor before she was moved from the old building, had delivered a healthy baby boy one hour after she arrived in the new hospital. The nurses had set up the new operating room to handle any emergency surgery. The nurses' aides had prepared the doctors' offices in readiness for the next day's clinic. The doctors had attended a few patients in emergency and admitted a young girl named Marieta. The chaplain had talked with Marieta and her father, and both had committed their life to Christ. Luz María and I were thrilled with the aides' report and looked forward to working in the new hospital.

As the day was ending, dark clouds hanging overhead abruptly gave way to a torrential downpour, but no one seemed to care. The "big move" was over. God had heard our prayers and provided good weather for our passage.

* * *

Next morning the nurses reported for duty at the new hospital. In contrast to the old building, the new facility was equipped with wall suction to aspirate excess secretions in postoperative patients, piped-in oxygen, two emergency rooms (one for contaminated cases), isolation units equipped to care for TB patients and other infectious cases, two pediatric rooms, where mothers could sleep on padded benches, and a central nursing station, where the nurse on duty had eye contact with the patients in intensive care, the nursery and the observation ward. The corridor at the nurses' station would no longer be cluttered with patients recovering from anesthesia and emergency trauma cases. The operating and recovery rooms were in a separate area, and admission to them was restricted to designated personnel. In the outpatient clinic, patients sat in a spacious waiting room with easy access to the chaplain's office. And the new hospital had a morgue! The expansion committee had designed a health-care facility that surpassed everyone's hopes.

Yet Hospital Vozandes-Shell was not quite finished. Construction workers still invaded the building. Stepladders cluttered the corridors, ceiling panels were missing, and workmen intent on installing window screens entered patients' rooms unannounced. The nurses' aides had no hot water with which to bathe their patients, and the nurses' call system that now replaced the old handbells was temperamental, sometimes working properly and sometimes not. Cartons of outpatient charts filled the center of the main waiting room. Staff, unable to file the charts until the carpenter finished the shelves, tried to keep children in the clinic from pulling them out of their boxes. These files were a constant temptation, and so was the fire-alarm button which was within easy reach of the older children. Its shrill sound brought staff running from every direction as only a few people knew how to turn it off.

In the operating room area, the plumbing functioned in reverse. When the nurses turned on the taps, water came up through the drain rather than from the spout. Toilets in the surgical unit overflowed, and the piped-in oxygen and nitrous oxide gases escaped before they reached the anesthetic machine. An undetected leak in the system put the surgical team through several stressful moments. Until the system could be adjusted to function properly, portable tanks were wheeled in to supply the anesthetic machine with the gases it required.

The hospital's official inauguration was only three weeks away. During those three weeks, staff entertained three different one-day tours and dealt with some unusual emergencies. The days passed quickly until, on the afternoon of March 30, 1985, some five hundred guests gathered on the hospital's front lawn for the new building's official opening. Most of the guests sat on benches borrowed from Luz del Evangelio church, in the shelter of a huge tent set up in case of rain, but bright blue skies and warm sunshine added to the beauty of the occasion.

During the inauguration ceremony, Hospital Administrator Jim Estes recognized the Swedish International Development Agency for providing funds to erect the new hospital. He also thanked the construction workers for their efforts and Ken Edgar and Wayne Wiley for directing and supervising the project. Dr. Eduardo Rodríguez, director of international relations in Ecuador, gave an address, recalling the excellent care that he had received in the old hospital in 1964 when he had been injured in a plane crash. Krister Gunnarsson of HCJB's Nordic Language Service brought a devotional message based on the verse, "Here is my servant whom I uphold" (Isaiah 42:1). In his message, he asked the audience, "What is the motivation of this work in Shell? What is the purpose of this new hospital?" Then he answered his own question by saying that as long as the hospital staff have the desire to serve God and humankind, their work will not be in vain, but will move forward. The ceremony ended when Christian Bausch, Sweden's ambassador, cut the ribbon and declared the hospital officially open.

Wearing their newest uniforms, hospital staff welcomed the visitors to the various departments. The tour of the building ended in the dining room and conference area where the women served refreshments and cut a six-foot long cake made to replicate the hospital. Two nurses had combined their artistic talents to decorate this masterpiece.

The celebration ended all too soon. After the chartered bus and private cars headed back to Quito, the men took down the tent and returned the church benches. The women cleaned the kitchen, and staff prepared supper for the inpatients.

On my way out of the hospital, I stopped in the large waiting room, now empty. On the wall, beside the entrance, hangs a plaque dedicating the new hospital to the glory of God and to the service of the people of Ecuador. A Scripture verse completes the inscription. It reads:

> *Therefore, my dear brothers, stand firm. Let nothing move you. Always give yourselves fully to the work of the Lord because you know that your labor in the Lord is not in vain (1 Corinthians 15:58).*

Beyond the parking lot a stony footpath winds through tall pastureland grass and ends abruptly at a footbridge. The 240-foot pedestrian bridge, suspended by steel cables anchored to cement towers, spans the deep ravine of the Motolo River. When the hospital project began, there was no easy access to the property. Anyone going to the new site took the long route into town, down a rain-rutted slope, and across a cracked cement bridge. Now the footbridge, built in the summer of 1984, links the mission's two properties and shortens the walking distance to the hospital for those living on the old site.

Suspended 75 feet above the water on that gently swaying bridge, I thought of the young man who had designed it. In the autumn of 1983, working visitors Mike and Ruth Ward had traveled from Johannesburg, South Africa, to Río de Janeiro in Brazil and on to Ecuador. During the three months they spent in Shell, Mike, a civil engineer, designed the footbridge, while Ruth, a fifth-year medical student, gained invaluable experience working at HCJB's jungle hospital.

Like Nate Saint, who never saw the fulfillment of his dream of a hospital serving the *oriente*, Mike never saw the bridge that he had designed. Perhaps he thought that he had contributed very little. Yet faithful service is never futile. Several years later, when the main bridge to the subdivision collapsed during heavy rains, the pedestrian bridge became the main link to the hospital for the many people arriving there on foot.

Except in the rainiest part of the year, the Motolo River appears to be a quiet, peaceful stream winding its way gently through the rain forest. Pausing now in the center of the bridge, I watched the flow of waters rippling across the stony riverbed and disappearing around the bend. On the vines and tropical foliage shrouding the steep riverbank, the late afternoon sun cast the shadow of a bridge and its lone occupant. I reflected on the purpose of the new hospital.

"The medical work performed here," I thought, "provides a link between us missionaries and the people of the rain forest. Just as this footbridge links the hospital with the people of this community, the health care that Hospital Vozandes-Shell provides is one of the Lord's bridges to people in need of physical and spiritual help. Uphold it, Lord, so that it may stand as long as it is needed!"

Bridge to the Rain Forest

Epilogue

Several years have passed since the inauguration of Hospital Vozandes-Shell. Yet still, on any given Monday morning, the large waiting room of the hospital's outpatient department overflows with clientele, many of whom have journeyed long distances to obtain medical help. The outpatient clinic, designed to accommodate the services of a maximum of four physicians, must now provide working space, as well, for Ecuadorian interns and residents who rotate to Shell from Hospital Vozandes-Quito's family-practice residency program.

Since tuberculosis is one of the leading causes of death in Ecuador, a dead-end corridor has been converted into a laboratory where highly trained personnel use sophisticated isolation techniques to detect the TB microbe. They also process specimens from other facilities within the country to help Ecuadorian medical personnel diagnose and treat their patients appropriately. Each month, on average, 10 new cases of TB are diagnosed at Hospital Vozandes-Shell.

The mission hospital maintains a unique relationship with the other evangelical organizations working in the area. The unity is such that few Indians and outsiders can separate the missions. For example, Joyce Stuck Grable of Gospel Missionary Union serves in the hospital as nurse chaplain. Having lived and worked in the jungle among the Shuar people since 1958, Joyce serves as translator between Shuar Indians patients and Spanish-speaking medical staff. While she counsels and comforts the seriously ill Shuar, two Ecuadorian chaplains minister to the spiritual needs of the hospital's Spanish-speaking patients, and when possible, staff do a follow-up ministry with home Bible studies. With the help of Shuar believers, Joyce has translated into their language the dialogue of the film, "*Jesus,*" a full-length dramatization of the entire Gospel of Luke.

Mission Aviation Fellowship pilots use "*Jesus*" extensively to evangelize jungle communities. From grassy airstrips in the jungle, these pilots also fly sick and injured people to Shell for the treatment of anything from a snake bite to a gunshot wound. Besides evacuating the ill, they fly community-service staff and their equipment to remote villages to further the continuing education of an increasing number of health promoters.

Brethren missionary Lloyd Rogers and his colleagues hold seminars for Christian Quichuas to prepare them to become leaders in their communities. He also arranges to send Ecuadorian young people to a seminary in Argentina for further training. Members of the Evangelical Missionary Association of Ecuador (AMEE) have been helping the Christian communities to obtain clear title to their land, an important step as oil explorations continue in the *oriente*.

Across the ravine from Hospital Vozandes-Shell, extensive renovations of the former Epp Memorial Hospital have transformed that building from a small medical facility into a quaint guesthouse. Former patients' rooms and surgery areas have now become guest rooms, a kitchen, a dining area, and a lounge. The Fuller Guesthouse, named in honor of Dr. Ev and Liz Fuller, can accommodate some 30 visitors. Photographs featuring staff and patients of earlier years, and a collection of memorabilia enliven its corridor for all to see.

Life at the jungle's edge has changed since those earlier days. The main thoroughfare, no longer a bumpy dirt road, is part of a paved highway that extends beyond Shell in both directions. The once sleepy town is no longer quiet and peaceful, for the noise of traffic continues throughout the nights. Between Shell and Baños, heavy road-construction machinery has cut across steep slopes, slicing off sharp curves to straighten the well-traveled highway. Yet with each improvement, the landscape in the Pastaza gorge loses something of its original beauty.

The eastern rain forest has changed, too, since the five American missionaries were killed by Auca spears. Their grave is gone, swept away by the angry Curaray River. Yet the territory holds the grave of another missionary, Rachel Saint. After an illness of several months, Rachel died in Quito on November 11, 1994. At her request, her body was buried among the Auca people whom she had served for many years.

A year before Rachel's death, Wycliffe missionaries completed a translation of the New Testament in the Waorani (Auca) language. No less significant has been the translation of the New Testament into the Pastaza Quichua language.

In June 1994 Waorani Indians stumbled on a piece of metal sticking out of the sand near the area that the five martyred missionaries had named "Palm Beach." The metal proved to be a part of Nate Saint's plane which, for 38 years, had been buried in the muddy river bottom. MAF pilot Bill Clapp flew to the area and identified the wreck. Later, Steve and Phil Saint, Nate's sons, along with Steve's son James and HCJB personnel, flew to the site to document the finding. The plane is now being restored as a memorial. Perhaps the Lord will use it to challenge others to serve Him in Ecuador's rain forest and all around the world.

Bridge to the Rain Forest

For Further Reading

Cook, Frank. *Seeds in the Wind.* Quito, Ecuador: World Radio Missionary Fellowship Inc., 1961.

Drown, Frank and Marie. *Mission to the Head Hunters.* New York, NY: Harper & Brothers, 1961.

Dufendach, Marilee (Ed.) *Catch the Vision.* Quito, Ecuador: World Radio Missionary Fellowship, Inc., 1989.

Elliot, Elisabeth. *Through Gates of Splendor.* New York, NY: Harper & Brothers, 1957.

Hitt, Russell T. *Jungle Pilot.* New York, NY: Harper & Brothers, 1959.

Landers, Kay (Ed.) *The Landscape of the Poor.* Devotional writings by HCJB missionaries. Quito, Ecuador: World Radio Missionary Fellowship, Inc.,1988.

Liefeld, Olive Fleming. *Unfolding Destinies.* Grand Rapids, MI: Zondervan Publishing House, 1990.

Neely, Lois. *Come Up to This Mountain.* Wheaton, IL: Tyndale Publishers, 1980.

Reichert, Richard P. *Daybreak Over Ecuador.* Quito, Ecuador: Sunrise Press, 1991.

Wallis, Ethel Emily. *The Dayuma Story.* New York, NY: Harper & Brothers, 1960.

__*Aucas Downriver.* New York, NY: Harper & Row, 1973.